PRAISE FOR *PACKAGING LOGISTICS*

"With the environmental problems posed by packaging waste now regularly in the news, the arrival of a book that puts packaging into a wider supply chain context is very welcome and timely. Henrik Pålsson, one of the main international specialists in the field of packaging logistics, has provided us with a comprehensive, well-researched and clearly written overview of the subject that should be widely read across the academic and practitioner communities."
Professor Alan McKinnon, Professor of Logistics, Kühne Logistics University, Hamburg

"This is a fascinating must-read for not just everyone working with packaging, but absolutely everyone in logistics and supply chain management, from raw material suppliers to wholesalers, retailers, and logistics service providers – and even consumers. Unlike other books in this area, there is no gap but an integration and reconciliation of marketing and supply chain efforts when considering packaging. It is a successful combination of a pragmatic, reader-friendly introduction to packaging logistics spiced with insights from the author's own cutting-edge research.

The book provides concrete tools for evaluating packaging performance throughout the supply chain, as well as for making crucial decisions with regards to technology and innovation, and the environment. It offers numerous highly interesting examples of the environmental implications of packaging: its interrelations with food waste, CO_2 emissions, to the energy footprint in the supply chain. The book provides a holistic view on the circular economy through its synergistic look at innovation and emission control in the supply chain.

Companies trying to expand to other markets, whether through an omnichannel approach, or to new markets in e.g. developing countries, will find the final chapters particularly useful. The book culminates in a comprehensive framework for the implications of packaging decisions across an e-commerce supply chain, and the final chapter widens the perspective to what to consider when introducing products in an emerging market."
Gyöngyi Kovács, Erkko Professor in Humanitarian Logistics, Supply Chain Management & Social Responsibility, Department of Marketing, Hanken School of Economics, Helsinki

"Makes a solid contribution to our understanding of the roles of packaging in supply chains."
Diana Twede, Professor Emeritus, School of Packaging, Michigan State University

"This book provides a comprehensive overview of contemporary topics in the field of packaging logistics. I find it useful for teaching as well as research purposes."
Erik Sandberg, Associate Professor in Logistics, Department of Management and Engineering, Divisions of Logistics and Quality Management, Linköping University, Linköping, Sweden

Packaging Logistics

Understanding and managing the economic and environmental impacts of packaging in supply chains

Henrik Pålsson

KoganPage

First published in Great Britain and the United States in 2018 by Kogan Page Limited

2nd Floor, 45 Gee Street
London
EC1V 3RS
United Kingdom

c/o Martin P Hill Consulting
122 W 27th Street
New York, NY 10001
USA

4737/23 Ansari Road
Daryaganj
New Delhi 110002
India

ISBN 978 0 7494 8170 4
E-ISBN 978 0 7494 8171 1

British Library Cataloguing-in-Publication Data

A CIP record for this book is available from the British Library.

Library of Congress Cataloging-in-Publication Data

Typeset by Integra Software Services Pvt. Ltd., Pondicherry
Print production managed by Jellyfish
Printed and bound in Great Britain by CPI Group (UK) Ltd, Croydon CR0 4YY

To Olivia and Arvid, my amazing and beloved children

CONTENTS

LIST OF FIGURES

LIST OF TABLES

ABBREVIATIONS

B2C	business-to-consumer
BCT	box compression test
CED	cumulated energy demand
DC	distribution centre
ECR	efficient consumer response
EOQ	economic order quantity
ICT	information and communication technology
ISO	international organization for standardization
KPI	key performance indicator
L (area in a packaging scorecard)	logistics
LCA	life cycle assessment
LDPE	low-density polyethylene
LTL	less-than-truckload
NFC	near-field communication
PM (area in a packaging scorecard)	packaging material
PW (area in a packaging scorecard)	product waste
QR	quick response
RFID	radio frequency identification
SRP	shelf-ready packaging
VA (area in a packaging scorecard)	value adding

PREFACE

Understanding and managing packaging can enable better-informed supply chain decisions, at both strategic and operational levels. This is because packaging has a major impact on supply chain performance, as it affects all logistics activities in supply chains. Packaging should protect and contain, unitize and apportion, communicate and offer convenience, as well as enable logistics and environmental efficiency. It is a complex matter to balance all these requirements, but if done successfully, it offers great potential for reducing costs and the negative environmental impacts in supply chains.

Packaging logistics is an emerging research field and a fast-growing area of interest for industry. This book combines the author's research with the synthesized findings of current research in the subject area and insights from industry. The content of the book has been shaped and developed over almost fifteen years of dedicated research and Master-level teaching on the subject. Throughout these years, and particularly in the teaching of Master's students in supply chain management and logistics programmes in Sweden and South Africa, there has been a lack of distinctly relevant literature on the subject. This book fills the gap. Its intention is to pedagogically present the theoretical basis, the fundamental concepts, and the research front of packaging logistics for Master's students and professionals in supply chain management, logistics, and packaging. The book is also useful for other educational programmes and professionals with an interest in packaging logistics.

The book covers the many roles of packaging in logistics throughout supply chains. It creates an in-depth understanding of packaging logistics by presenting fundamental ideas, concepts, principles, useful methods, organizational support and management support. It elaborates on the complexity of fulfilling many packaging requirements and balancing numerous trade-offs. It provides insights into the economic and environmental performance of packed products in supply chains, and it emphasizes a system perspective where the supply chain impacts of packed products are analysed.

ACKNOWLEDGEMENTS

In developing this book, I have benefited from the stimulating environment at Lund University where I work with supportive colleagues. Thank you all for being helpful and open to intriguing discussions. I am also grateful to all co-authors in academic papers. You have helped in developing ideas in different studies. I would like to acknowledge Henrik Wallström for his involvement in a research project on industrial packaging, where we refined and implemented a packaging evaluation model, which is presented in Chapters 8 and 9. I would also like to acknowledge all companies and authorities that have been involved in research studies. I particularly want to thank Eileen Deaner for proofreading and a number of researchers who have acted as reviewers for different chapters in the book: Carl Wänström, Erik Sandberg, Daniel Hellström, and Klas Hjort.

ABOUT THE AUTHOR

Dr Henrik Pålsson is Associate Professor in Packaging Logistics at Lund University where he undertakes research on a broad spectrum of logistics and supply chain management topics, primarily with a sustainability perspective. He has headed several long-term research projects within packaging logistics, e-commerce, and freight transportation. His research has been published in the *International Journal of Physical Distribution and Logistics Management*, *International Journal of Logistics: Research and Applications*, *Packaging Technology and Science*, *Journal of Cleaner Production* and in other peer-reviewed international publications. He has co-authored books, book chapters, and research reports, and he contributes regularly to international conferences, workshops and education. At Lund University, he is responsible for the Master's degree course and the PhD education programme in Packaging Logistics. He has international experience from teaching packaging logistics. Prior to entering academia in 2004, he worked with logistics management in industry.

Defining packaging logistics

01

Packaging is strategically important and has a significant impact on the logistics performance of supply chains. Informed decisions about packaging can reduce both supply chain costs and the environmental impact and increase the value of packed products. This book elaborates on packaging logistics, discusses the strategic role of packaging and its operational importance, presents useful frameworks and concepts, and provides useful decision support for packaging logistics.

Current research emphasizes that the key to obtaining cost-efficient packaging systems with minimal environmental impact across entire supply chains is to apply a systems approach and to utilize insights from supply chain integration. Balancing packaging requirements requires a focus on interactions and informed trade-offs between the many functions of packaging, such as protection, containment, apportionment, unitization, convenience and communication.

Packaging logistics is based on a considerable body of descriptive, explanatory and exploratory studies. It is an emerging research area and a fast-growing area of interest for industry. This book builds on the current research by synthesizing findings and explaining the theoretical basis of packaging logistics. It combines synthesized findings with insights from the author's own research, which forms the basis of the development of the novel ideas and frameworks explored in this book.

The book has the following outline. After defining packaging logistics in this chapter, the next three chapters emphasize its overall performance in supply chains from different perspectives: Chapter 2 focuses on packaging performance in supply chains, Chapter 3 provides a hands-on packaging performance methodology, and Chapter 4 focuses on the overall environmental performance of packaging logistics. Thereafter, Chapters 5 and 6 examine the different management considerations of packaging waste and of the information and communication technology (ICT) features of packaging. The next three chapters present various perspectives on industrial packaging: Chapter 7 presents characteristics of reusable packaging systems, while Chapters 8 and 9 discuss and illustrate how to evaluate the supply chain impacts of industrial packaging and how this can result in decision support. Chapters 10 and 11 cover organizational perspectives and concentrate on packaging development organization and packaging innovation capabilities. These perspectives are useful for packaging logistics because the way an organization is organized determines its ability to integrate various requirements into a packaging system solution, and its innovation capability for packaging affects its potential to address packaging logistics challenges in a novel way. The book ends with two chapters that discuss how to manage packaging logistics in two contexts with specific characteristics. The first context in Chapter 12 is the rapidly growing field of e-commerce where a number of contextual factors differ essentially from conventional trade. The second context in Chapter 13 is developing countries, which demonstrate a number of different characteristics to those considered in research studies to date.

Definition of packaging logistics

To define packaging logistics, we must first define packaging and the general levels of a packaging system, followed by a definition of logistics. The interactions between packaging and logistics are then discussed, leading to a precise definition of packaging logistics.

A well-accepted definition of packaging by Paine (1981) is that it is:

- a coordinated system of preparing goods for transport, distribution, storage, retailing, and end use;
- the means of ensuring safe delivery to the ultimate consumer in sound condition at minimum cost [and at minimum environmental impact];
- a techno-economic function aimed at minimizing costs of delivery while maximizing sales.

Packaging can be viewed as a system with three interrelated packaging levels (Figure 1.1). *Primary packaging* is closest to the product; it is often the consumer or sales packaging. *Secondary packaging* contains a certain

Figure 1.1 Three interrelated levels of a packaging system

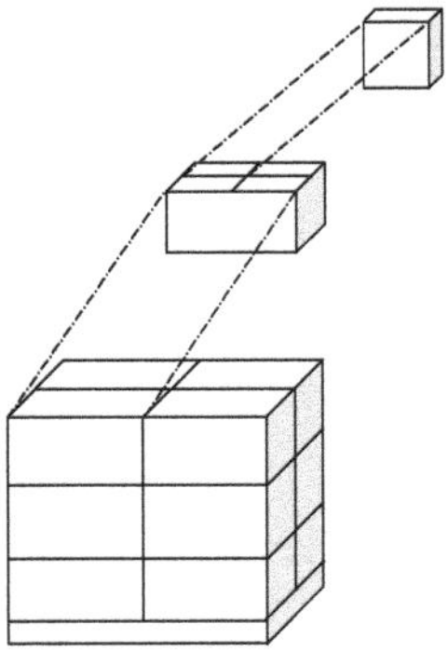

number of primary packages. Finally, *tertiary packaging*, such as a pallet or a roll container, contains several secondary packages.

The packaging system should fulfil six basic functions (Livingstone and Sparks, 1994; Lockamy III, 1995; Robertson, 1990):

- Protection: to safeguard the content.
- Containment: to hold and maintain the content.
- Apportionment: to reduce large-scale and high-volume production to manageable sizes.
- Unitization: to modularize the packaging levels to obtain material-handling and transport efficiency (eg to fit a number of primary packages into secondary packaging units, and a number of secondary packaging units into tertiary packaging).
- Communication: to identify the packaging in the supply chain and provide product information.
- Convenience: to simplify the use of products.

Throughout its life cycle, packaging is exposed to vibration, temperature variations, pressure and shocks. *Protection* means safeguarding the product or the contents of the packaging from these exposures. *Containment* is a related function that holds the contents and prevents them from interacting with the external environment. The package should be strong enough to hold the product or contents during the material-handling and transport processes in the supply chain. It should also avoid leakage, diffusion and permeation of the product, and the contents should maintain their characteristics until usage.

The packaging protection and containment for food products can be divided into three types of risks: physical, microbiological and chemical. Physical risks are foreign substances, such as glass, metal and plastics. Microbiological risks, such as bacteria, viruses and fungi, depend on time and temperature in the supply chain. Chemical risks can be allergens, overdoses of preservatives, and traces of heavy alloys and pesticides.

To obtain economies of scale, the output of production is often high volumes. In order to transform such large-scale and high-volume production to manageable sizes, packaging is used to apportion the products. An example of *apportionment* is when fruit juice from tanks is packed in 33 cl bottles and 1-litre juice packages. Spinach is another example; it is sold in smaller packages than it used to be, which means the consumer can defrost smaller amounts. This reduces food waste, as spinach should not be refrozen.

Unitization facilitates logistics and material handling. Through modularization of the packaging levels, a number of different primary packages can be grouped together to fit in secondary packaging, and a number of different types of secondary packages can be grouped together to fit in tertiary ones. In so doing, the overall number of packaging units that are handled is minimized. In a picking process, for instance, one secondary packaging unit containing several primary packaging units can be handled, and in the loading and unloading process for transportation, one pallet with several secondary packaging units can be handled. Unitization also facilitates repacking in a warehouse from a pallet dedicated to one product into a mixed pallet, as a number of secondary packaging units instead of individual primary packaging units need to be picked.

Packaging has a *communication* function from two perspectives. First, it is key in providing logistical track & trace data through labelling throughout the supply chain. Labels with barcodes, radio-frequency identification (RFID) or other identification technologies in manufacturing can be used to determine the location, time, temperature and other pieces of information related to the packed product for warehouses, ports, transportation, retail stores, etc. Second, the packaging system also communicates with consumers. It helps them to identify products through branding, design, shape and labelling as well as providing product information, such as nutritional content and how the product has been manufactured.

Finally, packaging should provide *convenience* for users by simplifying the use of packaging and the products within them. This includes ensuring that packaging units are easy to open, close, handle and dispose of throughout the supply chain. Examples are handles that simplify consumer handling, dispensing caps, portion indicators (eg in rice packaging) and a serving bowl (eg ready-cooked food).

In addition to the six main functions, packaging affects the logistics and environmental efficiency in supply chains because it interacts with material-handling equipment, information systems, manual handling, transportation and waste management. From a logistics perspective, packaging affects and is affected by every logistics activity, such as material handling, storage, transportation and waste handling (Ballou, 2004; Bowersox *et al*, 2002). For instance, both the shape and the dimension of packaging affect cubic utilization efficiency in transport; the choice of material influences waste management and recycling; and the protectiveness of the packaging affects the amount of product waste in the supply chain.

Now that packaging has been defined, let us consider logistics to complete the definition of packaging logistics.

According to the Council of Supply Chain Management Professionals, logistics management is 'that part of SCM that plans, implements and controls the efficient, effective forward and reverse flow and storage of goods, services and related information between the point of origin and the point of consumption in order to meet customers' requirements' (CSCMP, 2018).

By combining the two definitions of packaging and logistics, it is obvious that packaging affects every logistics activity from point of filling to point of consumption. Thus, in this book, the term 'packaging logistics' emphasizes the interface between packaging and logistics throughout supply chains, which demonstrates and supports the importance of viewing the physical flow of goods and its related information flow as one integrated system.

The three generic levels of a packaging system (primary, secondary, tertiary) were defined at the beginning of this section. The current literature and practice also use alternative names, which are summarized in Table 1.1. Special notice should be given to differences between consumer packaging and industrial packaging. Consumer packaging includes marketing aspects and should take into account any legislative requirements related to consumers, whereas industrial packaging can focus on other packaging functions, such as providing logistics efficiency.

The definition of each packaging type highlights its main characteristics. However, to understand the characteristics of shelf-ready packaging (SRP), additional pieces of information are provided. SRP consists of packaging that is delivered to a retailer in a ready-to-sell merchandised unit. It can be placed directly on the shelf without repacking or unpacking (other than taking off an outer layer of plastic film or perforated cardboard, for example). SRP can consist of a tray on a shelf, a merchandise unit on

Table 1.1 Definitions of different packaging types

Packaging type and definition
Primary packaging, consumer packaging, sales packaging
Packaging which is in contact with the product, and the one that consumers usually bring home
Secondary packaging
Contains a certain number of primary packages
Tertiary packaging
Contains a certain number of secondary packages
Transport packaging, industrial packaging, distribution packaging, bulk packaging
Packaging that facilitates handling, transport and storage of a number of primary packages in order to provide efficient production and distribution, as well as to prevent physical handling and damage during transportation
Group packaging
Packaging that is created to facilitate protection, display, handling, and/or transportation of a number of primary packages
Display packaging
Same as group packaging, often with an emphasis on display features
Shelf-ready packaging
Same as group packaging with a special emphasis on the design to fit in retail stores
Used packaging
Packaging or packaging material remaining after the removal of the product it contained

the floor (often for promotional display), and reusable plastic trays (see examples in Figure 1.2). The functional requirements of SRP are that it should be easy to identify, open, replenish, dispose of, and purchase (ECR Europe, 2007).

SRP has a number of functional requirements. The product should be visible through the packaging, the product information should be clearly displayed, and the label easily accessible. The brand should face the customers, but the logistics data should not. SRP should have simple instructions, preferably with pictures rather than text to avoid

Figure 1.2 Examples of shelf-ready packaging (SRP)

language barriers. The quality and appearance of the packaging should be unchanged after opening. To facilitate replenishment, it is essential that the SRP is stable, with the primary packages in the right position after unwanted packaging material is removed. A similar requirement is that the primary packaging should be easy to pick without affecting the SRP or the primary packaging negatively after being picked.

The levels of a packaging system

As noted above, packaging should be described as a system with three interacting packaging levels. This emphasizes the fact that the performance of a packaging system depends on the performance of each individual packaging level, on the interactions between these levels, and their interactions with the product (see Figure 1.3). For instance, at each packaging level it should be possible to identify the product without considering the intermediate packaging levels: eg temperature requirements should be visible on both the secondary and tertiary packaging. Another example is that the product might reduce the strength of the packaging needed: a tightly packed engine block can support stackability, because the product itself is strong, thus requiring only minor packaging. The usual interactions of the six basic functions of the packaging system are described in Table 1.2.

Packaging affects the whole supply chain, from point of filling to point of consumption. Figure 1.4 shows the principal and simplified supply chain for a packed product, in which a minimum number of process stages are highlighted. Each stage involves packaging requirements, such as adaptation to logistics equipment, stackability, protection, maximum weight, labelling

Figure 1.3 Illustration of the packaging system for the packed product

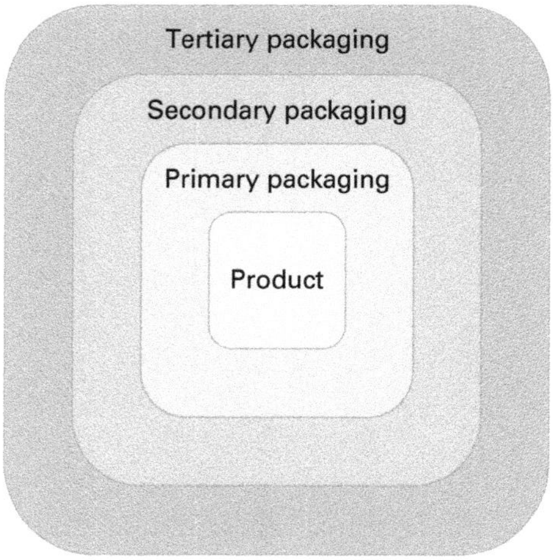

Table 1.2 Usual interactions of the six basic functions of the packaging system

Packaging function	Interactions	Example
Protection	Each packaging level and the product interact for total packaging system protection.	A strong product can minimize the need for packaging strength, while a pressure-sensitive product may require extra packaging. Additional secondary packaging may reduce the need for tertiary packaging and vice versa.
Containment	To prevent the content from interacting with the external environment, consider the combined containment of the levels of the packaging system.	If the primary packaging contains the product sufficiently even when it is handled in secondary packaging, there is no need to add further support. However, if the external environment is more challenging where secondary packaging is used than when individual primary packaging is handled, the additional support should be added to secondary packaging.

(*continued*)

Table 1.2 *(Continued)*

Packaging function	Interactions	Example
Apportionment	Handling each packaging level should be manageable where it is used in the supply chain. The size of the primary packaging should be aligned with consumer needs, while secondary packaging should be manageable for repacking in a distribution centre and in replenishment in a store, etc. The addition of each extra primary packaging unit adds a discrete weight and volume to the secondary packaging. Thus, the apportionment (weight and volume) of each packaging level interacts with the other levels.	2-litre packs of ice cream are sold to families because they are manageable and have long shelf life. For efficiency, 6 big packs (12 litres) are handled in a secondary packaging unit in the distribution centres and in replenishment, which is just below the maximum weight for ergonomic handling. With 2.5-litre packs, only 4 packages (10 litres) can be handled in a secondary packaging unit, which would require more handling.
Unitization	The number of primary packaging units should fit the secondary packaging unit, and the number of secondary units should fit the tertiary packaging unit.	20 milk packages with a quadratic base (74 mm) fit a secondary packaging (300 × 400) unit. Eight of these fit on a Euro-pallet. The secondary packaging units are handled to obtain mixed pallets in a warehouse.
Communication	The labels at each packaging level should inform the supply chain actors about the product and its requirements. The secondary packaging should be linked to the primary, and the tertiary to the secondary.	Scanning the tertiary packaging should provide product information and the number of secondary packages in the tertiary packaging.
Convenience	Convenience is associated with consumers, who mainly handle primary packaging and occasionally secondary packaging. Interactions are mainly related to multipacks where a carrying capacity in the secondary packaging eliminates such a need in the primary packaging.	Bottles are often grouped and sold in multipacks (eg 6, 10 or 24) with a handle to make them easy to carry.

Figure 1.4 A principal and simplified supply chain for a packed product

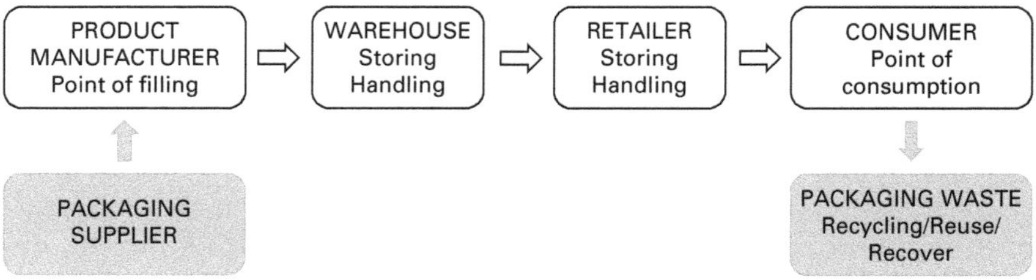

Figure 1.5 A simplified supply chain for a packed product

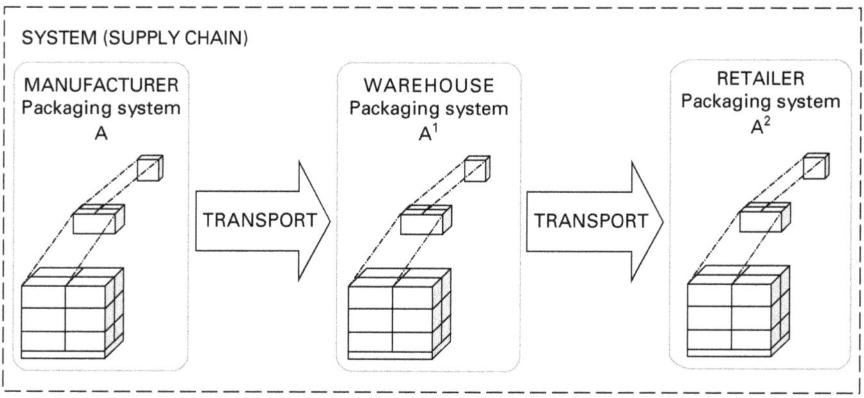

and communication. Between each stage, the packages are transported with requirements on volume and weight efficiency as well as adaptability to other packaging for co-loading.

Figure 1.5 further describes the packaging system in each stage. The packaging system is completed at the manufacturer, and then transported to a warehouse. There, it is modified during the picking process, which is indicated by A^1. This modified packaging system is transported to the retailer, where it can be modified again or kept unchanged, indicated by A^2. To address and match all packaging requirements from different actors in the supply chain (see Figure 1.4) as well as interactions and changes in the packaging system (Figure 1.5), a systems approach needs to be applied.

Systems approach in packaging logistics

A system is a group of individual parts that work together to form a unified whole. Any business, organization or supply chain can be studied as a system. It is a way to examine a phenomenon or linked parts as one unit, instead of as individual units. A system is characterized by the whole system not having to be equal to the sum of its parts, as these parts are connected to each other to form a whole. An essential characteristic of this approach is the focus on the interactions between the parts. Thus, the 'behaviour of a single autonomous element is different from its behaviour when the element interacts with other elements' (Mele *et al*, 2010, p 127). Another essential characteristic is that a system contains a clear system boundary. However, a system can be part of a bigger system or overlap with another system. A packaging system, for instance, can be part of a supply chain, which is a bigger system. In this case, the packaging system can be considered as a sub-system.

Another basic characteristic is the distinction between open and closed systems. In open systems (Figure 1.6), exchanges with the surrounding environment need to be considered, while in closed ones there are no exchanges. As an example, a reusable packaging system in a closed loop would be considered a closed system, while a one-way packaging system in a supply chain from point of filling to point of sale is an open system. The surrounding environment that affects the system studied would consist of the packaging supplier, the consumer and the end of life of the packaging.

A systems approach helps us to understand how all the parts influence each other within the bigger system. It facilitates problem-solving, by viewing problems and challenges as parts of an overall system, rather than reacting to specific parts or outcomes, and helps avoid sub-optimization by focusing on the primary purpose of a system. By examining a system as a whole, it is easier to understand how each part contributes to the overall purpose. Key characteristics of the systems approach are that it facilitates the pinpointing of trade-offs and interactions between the parts in the system.

Figure 1.6 A principal open system with nodes and links

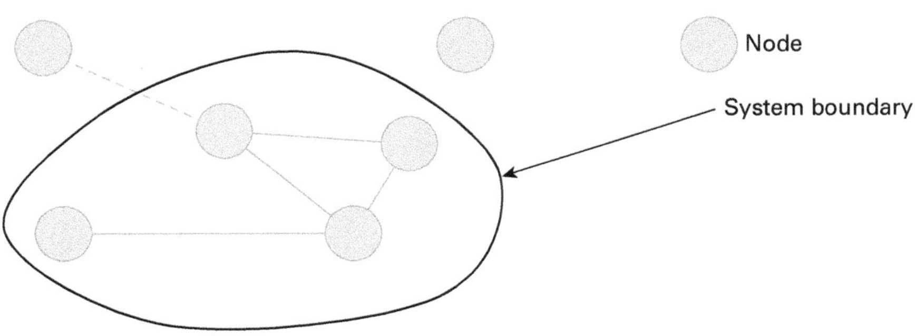

Both trade-offs and interactions are fundamental in understanding packaging logistics. For instance, adding perforation to a package could increase consumer convenience, but this could also decrease package stackability in warehouses and during transportation. Thus, there is a trade-off between consumer convenience and supply chain efficiency. In terms of interactions, packaging systems interact with logistics equipment throughout supply chains. An example is a conveyor belt designed to handle one particular set of packaging dimensions that must be taken into account in the selection of the ultimate packaging.

When applying a systems approach to investigate how the individual parts in a system influence each other, a powerful technique is to *zoom in* and *zoom out* repeatedly. Zooming in means that one part of a system is enlarged to provide a detailed view of that particular part, while zooming out shows each part's links to and overall influence on the system (Figure 1.7). By continuously shifting between zooming in and zooming out, the effect of a modification to one part on the whole system can be identified, thus avoiding sub-optimization. Chapter 3 describes how the idea of zooming in and zooming out can be applied in the analysis phase of the packaging performance methodology.

In short, a systems approach in packaging logistics enables us to obtain an overall picture of the many roles of packaging. This helps us to analyse and balance packaging system requirements and trade-offs, and to highlight interactions. By balancing these requirements, the systems approach enables us to focus on the total requirements for the packaging system. Trade-offs and interactions are found within the packaging system (primary, secondary,

Figure 1.7 Level of enlargement in systems by zooming in and out

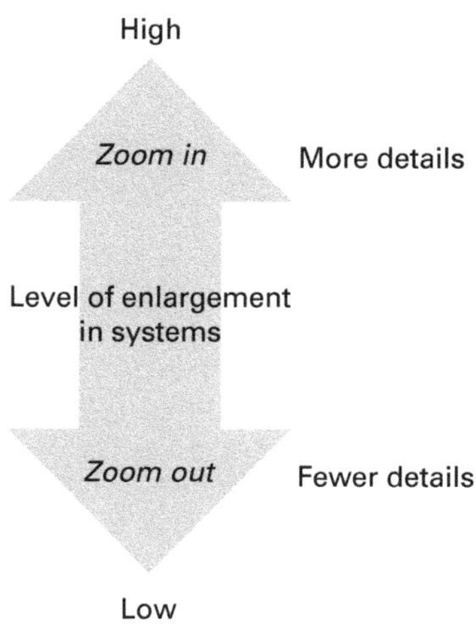

tertiary), between functional areas (logistics, marketing, environment, etc) and between supply chain actors. As the systems approach helps to balance requirements in a structured way, it also helps to communicate this to actors in functional areas and in the supply chain in general.

The role of supply chain integration for packaging logistics

In addition to a systems approach, *supply chain integration* is a key concept for packaging logistics. The term refers to coordinating and integrating separate parts of a supply chain into a cohesive whole. The level of integration depends on 'the extent to which separate parties work together in a cooperative manner to arrive at mutually acceptable outcomes' (O'Leary-Kelly and Flores, 2002). The underlying logic for supply chain integration is to synchronize all processes and activities with a focus on the end-customer offer. Supply chain integration aims to optimize flows rather than activities. In doing so, it balances costs and services throughout the supply chain.

The flows to be integrated are either physical or information. Packaging plays a vital role in both. In the physical flow, packaging influences logistics from the point of filling until the point of consumption. This includes the necessary synchronization of requirements on packaging functions (eg protection, handling and stackability) from many actors. In the information flow, packaging is an information carrier throughout the supply chain, containing both logistical data for track & trace capabilities and product information. Packaging should enable communication with all actors through identification technologies and labelling. Accordingly, packaging systems should be designed or selected to facilitate the integration of both physical and information flows.

Integration in a supply chain can take different forms. Figure 1.8 shows two basic types of integration: vertical and horizontal. Vertical refers to integration with suppliers and customers. Based on a certain company, this can also be divided into backward and forward integration. Horizontal refers to integration between competitors. Two other categories are internal and external integration: internal refers to integration between functions within a company, whereas external is less distinct than vertical and horizontal as it refers to integration between companies. Packaging affects each type of integration. Table 1.3 presents a number of examples.

The level of supply chain integration can be described in four progressive stages (Stevens, 1989). In the first stage, companies are functionally driven

Figure 1.8 Illustration of different types of integration

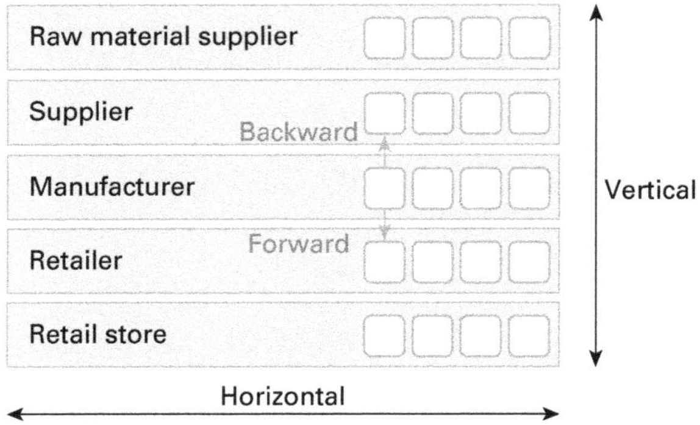

Table 1.3 Packaging examples for different types of integration

Integration type	Description	Example of packaging integration
Vertical	With suppliers and customers	Combine packaging requirements on handling, stackability and communication etc from various companies
Horizontal	Between competitors	Modularization of packaging within an industry for co-loading
Internal	Between functions in a company	Marketing and logistics requirement on packaging Concurrent product and packaging design
External	Between companies	See vertical and horizontal integration
Backward	Upstream in the supply chain	See vertical integration
Forward	Downstream in the supply chain	See vertical integration

with no integration. Here, packaging is usually seen as a commodity with a focus on minimizing purchase cost.

In the second stage, companies are functionally integrated, meaning that the organizational units related to a given function, such as logistics management or manufacturing, are integrated. However, companies in this stage still have discrete business functions, often divided by inventory. In terms of packaging, there is still a focus on cost, but the 'ownership' and location of the packaging activity within the organization may affect this. For instance, if it comes under internal logistics, the packaging focus tends to be on fulfilling internal material-handling requirements. If it comes under marketing, the focus tends to be on commercialization.

In the third stage, companies become internally integrated, with visibility of the goods flow from purchasing to distribution. Here, companies tend to focus on internal efficiency, which means that trade-offs between internal packaging requirements are considered.

In the last stage, where companies have external integration, the focus is on effectiveness to ensure that the right things are done. External integration involves being customer-orientated rather than product-orientated, focusing on cooperation with suppliers rather than price negotiations. Here, packaging decisions based on informed trade-offs between supply chain requirements are possible. This depends on visibility, transparency and collaboration in the supply chain. To be able to make these types of decisions, methods and tools for packaging selection need to support such an approach.

Other challenges for an integrated approach to packaging are conflicting goals and attitudes in functions or companies. An integrated approach to packaging selection also means trade-offs, where one organization might benefit while another may have an internal disadvantage. Thus, cost and benefit sharing within the supply chain is essential to obtain a sustainable integrated approach, but this is often a challenge. In addition, integration requires information sharing, but this also raises a number of challenges. In particular, companies tend to have different IT systems that are not always compatible; information sharing can be costly, and organizations see a risk of losing a competitive advantage by sharing information. Other challenges are related to the complexity of having a large supplier base or many customers. Then there is a need to develop a strategy for identifying how to capture, structure and prioritize packaging requirements from the majority of organizations. Finally, there may be physical barriers to integration in the material flow. A company, for instance, may not be able to handle certain packages due to limitations in its equipment.

The systems approach and supply chain integration are essential to packaging logistics. The logic and characteristics of both have similarities but are also complementary. Both emphasize a holistic view to avoid sub-optimization as well as to coordinate and synchronize requirements and parts in the system or supply chain. The systems approach is flexible in defining system boundaries and it can include sub-systems, for example, a packaging system can be a sub-system within a supply chain, which can be considered as the main system, while the boundaries for supply chain integration, on the other hand, are related to the organizations included. For packaging logistics, the systems approach offers a systematic way to address the complex interactions of packaging systems within supply chains from a holistic perspective, while supply chain integration offers ideas and theoretical support for integrating packaging-related activities and actors in supply chains.

Summary

This chapter pointed out that packaging is strategically important and has significant impacts on logistics performance in supply chains. Informed decisions about packaging can reduce both supply chain costs and environmental impact, and increase the value of packed products. Packaging should be viewed as a system with three interrelated packaging levels: primary, secondary and tertiary. Complementary terms for these packaging levels were described. The basic functions of packaging are to protect, contain, apportion, unitize, communicate and provide convenience. The communication function is key in providing logistical track & trace data by means of labelling throughout the supply chain, as well as in reaching out to consumers. The six functions should be combined with features that facilitate logistics and environmental efficiency in supply chains, since packaging interacts with material-handling equipment, information systems, manual handling, transportation and waste management.

The chapter also elaborated on packaging logistics by defining and discussing packaging and logistics. It emphasized that the performance of a packaging system in a supply chain depends on the performance of each individual packaging level, on the interactions between the packaging levels, and on their interactions with the product.

It was highlighted that a systems approach and supply chain integration are essential to packaging logistics. Their logic and characteristics have similarities, but they are also complementary. Both emphasize a holistic view to avoid sub-optimization as well as to coordinate and synchronize requirements and parts in the system or supply chain. A systems approach is useful to understand and analyse the performance of a packaging system because it helps to understand how the parts influence each other from within to present an overall picture of the many roles of packaging. A systems approach helps to identify and consider trade-offs and interactions in packaging logistics.

In order to apply a systems approach in practice and to understand how its parts influence each other within the system, a powerful technique is to zoom in and out repeatedly. Supply chain integration involves coordinating and integrating the separate parts of a supply chain into a cohesive whole. Packaging plays a vital role in the integration of both the physical and information flows. In the physical flow, packaging affects every logistics activity from the point of filling until the point of consumption. In the information flow, packaging is an information carrier throughout the supply chain containing both logistical data for track & trace capabilities and product information. The integrations can be vertical, horizontal, internal, external, backward and forward. The chapter presented examples of packaging to illustrate its role in these different types of integration.

Packaging performance in supply chains

02

Chapter 2 highlights and discusses the role of packaging in adding value in logistics throughout the whole supply chain. A key argument is to apply a holistic view to the supply chain thus making informed trade-offs between various packaging requirements. The chapter also discusses standardization and packaging postponement, and their impacts on packaging performance.

Packaging as a value generator and enabler of cost efficiency

Traditionally, packaging is often viewed as a 'necessary evil' that generates waste. Research and practice, however, have demonstrated that packaging is strategically important and has a significant impact on supply chain performance. A supply chain approach to packaging can generate value and enable cost efficiency by reducing product waste and improving logistics and transport efficiency. Packaging generates value in sales by attracting customers using graphical and other design features, and by providing convenience, such as facilitating the handling of the packed product. Packaging can reduce product waste by offering protection against physical and chemical effects, temperature changes, insects and more. It can also apportion products according to stakeholder needs, thus reducing product waste in the supply chain and in the consumption phase. Information displayed on the packaging about track & trace data and on how to handle and consume the product also reduces product waste because of better control, fewer mistakes and reduced theft. When it comes to logistics and transport efficiency, packaging affects warehouse and transport utilization as well as production and handling efficiency. A more detailed discussion on reduced product waste and on logistics and transport efficiency is presented in Chapter 4.

To maximize value generation and the total cost efficiency of packaging requires a holistic approach to developing and selecting packaging systems where the system boundaries include the whole supply chain. A holistic approach to the supply chain highlights the need to manage trade-offs between requirements on packaging from actors in the supply chain.

Key issues in evaluating supply chain performance of packaging

Trade-offs between various packaging features must be considered in evaluating supply chain performance of packaging systems. Ideally, packaging should fulfil requirements from various areas, such as logistics (including warehousing, handling and distribution), marketing, production, sourcing and consumers. Packaging should also have a minimal environmental impact and fulfil legal requirements. However, these different requirements may be in conflict. For instance, actors often have specific requirements for protection, apportionment, and track & trace capabilities, but then manufacturing often emphasizes filling efficiency and low packaging cost. The main warehouse requirements may be stackability and handleability for efficient warehouse operations. Transport, on the other hand, may focus on handleability as well as volume and weight efficiency. The retailer may accentuate promotional attributes, product information and handleability for sufficient replenishment. Consumers often appreciate convenience and apportionment of packaging. Marketing requirements for packaging size may conflict with volume-efficient packaging requirements in logistics. These examples illustrate the complexity of evaluating the supply chain performance of packaging systems and being able to make informed trade-offs between packaging features for different actors. To address these potential conflicts, the design of a packaging system needs to address a number of trade-offs.

The trade-offs in a packaging system depend on two types of requirements that come into conflict: those from the various areas (eg logistics, marketing, production, sourcing and consumers) related to the economic and the environmental impact, or those from different actors in the supply chain and other stakeholders (eg legislation). These different requirements may overlap or be contradictory. For instance, a manufacturer may aim to source low-priced packaging material, while the retailer prefers a more expensive package to promote sales. At the same time, the same packaging feature may

Table 2.1 Examples of common packaging system trade-offs

	Logistics	Marketing	Production	Sourcing	Consumer
Packaging size	Volume efficient Easy to handle	Visible (sometimes large is preferable)	Volume efficient Easy to handle	Unitization	Convenient
Packaging labelling (logistics information)	Identifiable	Graphically clean with minimal labels	Identifiable	–	Graphically clean with minimal labels
Packaging material	Low weight, but proper protection	Promote sales	Low weight, but proper protection Avoid dust in handling	–	Can be recycled
Packaging cost	Protect properly	Promote sales	Context specific	Low price	Context specific
Standardization	Medium no. of alternatives Handling efficiency Stackability Transport efficiency Co-loading	Aim for unique packaging – many alternatives	Medium no. of alternatives Handling efficiency Space utilization	Few alternatives	Context specific

have different requirements from different actors in the supply chain: the manufacturer may prefer certain dimensions for internal material handling, while the distribution centre needs different dimensions for their material handling. Table 2.1 presents common packaging system trade-offs.

In Table 2.1, *packaging size* refers to the packaging dimensions. In logistics and production, the focus is on cubic (volume) utilization and ensuring the packaging systems are easy to handle, whereas marketing tends to focus on visibility, which sometimes equates to large packaging. From a sourcing perspective, the size should be unitized with other packages to obtain economies of scale in purchasing, while consumers typically focus on the convenience of the packaging size. *Packaging labelling* should enable logistics and production to identify the packed products. This can decrease shipment delays and lost shipments, and reduce the time for order fulfilment in warehouses. However, marketing and consumers usually prefer packaging which has minimal labels with logistics information. *Standardization* can enable logistics efficiency in various ways, such as efficiency in loading and unloading, and facilitate stackability, but it can have a negative effect on sales, where unique alternatives may be needed. (More discussion on standardization is presented in the next section.) The examples of trade-offs here are just a few of the potential ones. More trade-offs will be highlighted throughout the book.

Standardization

Standardization can facilitate efficiency in supply chains, but it can also result in 'lock-in'. For instance, a standardized packaging assortment may fit current products, vehicles and material-handling equipment perfectly. However, it may lead to inefficient handling and vehicle utilization over time if the characteristics of the product assortment or the supply chain change. It may simply be too expensive to invest in a new packaging assortment. Thus, the risk of a lock-in effect should be included in the decision to standardize.

In general, the International Organization for Standardization (ISO) states that standardization should reduce complexity, protect the interests of majorities, target selected activities, persist over time, but be subject to review and change when needed (Min et al., 2014). The main aim of logistics standardization is to become more efficient by eliminating waste, duplication of work and redundant activities. A low level of logistics and packaging standardization often results in greater logistics cost. Logistics standardization, which is affected by packaging, can be grouped into four areas: packaging (eg size and materials), transportation, material handling (storage, and loading and unloading), and ICT (Min *et al*, 2014).

Packaging standardization refers to the dimensions of packaging and load carriers, such as pallets and containers. This can help to maximize cubic utilization of vehicles (eg trucks, trains and aircrafts), facilitate co-loading of unitized packaging, and simplify material handling. Standardized packaging can also facilitate the integration of physical flows, both intra- and inter-organizationally. In Europe, the Euro-pallet of 1,200 x 800 mm is an example of a standardized load carrier, which provides these kinds of opportunities. Other common ISO standard pallets are presented in Table 2.2. In international shipping, examples are 20 ft and 40 ft containers. Another kind of packaging standardization is related to the packaging material. Such a standard can describe which materials to use and provide material quality guidelines. Packaging standardization can also refer to weight restrictions that ensure ergonomic handling of packaging and that the weight restrictions of logistics equipment are met. Finally, the standardization of packaging can include recycling guidelines for packaging waste to obtain high levels of recycling.

Transportation standardization refers to the dimensions of freight containers and rail cars to fit unit loads. It should facilitate the matching of vehicle size with packaging systems in order to accommodate as many packed products as possible. Within the EU member states, for example, a modular system for trucks is used. The limitations for dimensions and weight started to become harmonized in the 1970s. Currently, the EU limitations are 18.75 metres length, 2.55 metres width, 4.0 metres height and 40 tonnes weight. Some countries, such as Sweden and Finland, have exceptions for domestic traffic. By using these standardized dimensions, it is easier to carry out international transportation and to use modularized vehicles for flexible transportation between rural and urban areas without unloading and reloading packages.

Table 2.2 International pallet standards of ISO

Dimension (mm)	Description
1200 × 800	European standard pallet
1200 × 1000	Standard pallet for Germany and the Netherlands (complement to the Euro-pallet)
1140 × 1140	Good fit with sea containers
1100 × 1100	Common in Far East Asia
1219 × 1016	US standard pallet (for domestic use)

Material-handling standardization includes storage and equipment standardization. The latter refers to unified material-handling equipment, such as automated storage, racks, conveyor belts and forklift trucks. This facilitates effective and efficient material-handling operations, including loading and unloading throughout supply chains. To achieve efficiency, the standardization of material handling should be aligned with that of packaging. For instance, if reusable packaging is automatically handled in a conveyor belt, the dimensions and weight restrictions, the labels with logistics information, etc, of the packaging must function in this material-handling equipment. Another example is forklifts that must be adapted to pallets, so that the gap between the forks is designed to lift Euro-pallets, if these are the main packaging systems to be handled. The storage aspect of material-handling standardization also aims to maximize the storage area by unifying storage equipment. This aids mixed and flexible storage points.

ICT standardization can help actors in the supply chain to integrate logistics information systems, which, in turn, can help to coordinate physical material and delivery processes. Packaging plays a key role in obtaining logistics efficiency from ICT (see Chapter 6). By accessing track & trace data about packaging systems, vehicle positions, loading and unloading activities and so on, companies in a supply chain can act as a unit. This requires ICT standardization of identification technologies, such as barcodes and RFID technology, their communication linkages (eg electronic data interchange (EDI)), and ICT interfaces. It also requires software with capabilities to integrate information without disruption from companies in numerous supply chains.

Packaging postponement

Common trade-offs occur between logistics and marketing, as Table 2.1 shows. This can relate to, for example, packaging size or number of packaging types. Packaging postponement is a possible strategy to manage some of these trade-offs. As Twede, Clarke and Tait (2000) describe, packaging postponement is useful for customization to local markets of globally distributed products.

Postponement is well established in logistics. It is based on the idea that by postponing activities until more information is available or when diversity is needed, the earlier stages of a supply chain can be more efficient. Postponement can be obtained by delaying logistics activities or manufacturing and packaging activities. Four types of postponement are described in Table 2.3. Packaging affects all four, which are discussed below (based on Twede, Clarke and Tait, 2000).

Table 2.3 Role of packaging in postponement and speculation strategies

Strategy	Explanation	Effects on packaging
Full postponement	Postponement in manufacturing and logistics	Bulk transportation Customized packaging vs packaging production efficiency and inventory costs
Manufacturing postponement	Postponement in manufacturing; speculation in logistics	Bulk transportation Customized packaging vs packaging production efficiency or inventory costs
Logistics postponement	Speculation in manufacturing; postponement in logistics	Standardized dimensions for the packaging assortment vs volume and weight efficiency Strong to enable handling and stacking requirements
Full speculation	Speculation in manufacturing and logistics	High-volume packaging Multilanguage labelling

SOURCE Based on Pagh and Cooper, 1998

Full postponement refers to the last step of the manufacturing process, as well as logistics and packaging being postponed until a customer order is received. Semi-finished products are stored in a central warehouse and customized, packed and shipped once an order is placed. For instance, standardized white T-shirts for the European market can be manufactured and stored centrally. Once a customer order is received and before it is shipped, the T-shirt is dyed and combined with customized items and customized packaging, such as labelling, country-specific instructions and so on.

Manufacturing postponement utilizes production speculation and economies of scale to some extent. Semi-finished standardized products are manufactured and shipped in bulk to a warehouse close to the market. The packed products are finalized in the warehouse once a customer order or a short-term forecast is received. Manufacturing postponement increases transport efficiency, reduces the risk of obsolescence, but increases the costs of production and packaging. Three examples of manufacturing postponement are:

- The distribution of unassembled products to consumers, which enables products to be sold in volume-efficient packages. The assembly process is postponed until the consumer stage of the supply chain. This enables volume-efficient transportation and material handling throughout the supply chain. Packaging is an essential part of this approach because it has to facilitate volume efficiency, stackability and handleability as well as provide assembly information.

- Product, eg computers, that can be stored centrally and provided with country-specific packaging once an order is received.
- Customization of the number of items ordered: by postponing packaging, the package can be customized to fit the number of items ordered. Empty space and extra filling material can thereby be avoided. Recent developments in creating on-demand packaging has rapidly facilitated this type of packaging postponement. It should be noted, though, that packaging production in general benefits from economies of scale.

Logistics postponement utilizes production speculation and economies of scale, but the logistics activities are postponed until an order is placed. This reduces the capital cost of inventory and the risk of obsolescence due to the maintenance of lower levels of total safety stock in a few centralized warehouses, rather than having safety stock in several decentralized warehouses, as in full speculation. However, transportation costs are increased due to the faster delivery speed. This leads to more less-than-truckloads (less than load/LTL) and more costly modes of transport, such as air freight. To support logistics postponement, packaging should have standardized dimensions and labels to facilitate automatic sorting and picking, but still be volume and weight efficient for the LTL transport. The packages are typically handled in complex supply chains and in a variety of supply chains, which means that they need to be strong to withstand handling and stacking requirements.

Full speculation refers to mass production focused on economies of scale in production. Usually, full truckloads of finished goods are transported to warehouses close to the expected demand. Full speculation means that packaging is standardized with large volumes. It commonly serves several markets, which requires multilanguage labelling. The downsides of full speculation are relatively high inventory costs and the risk of obsolescence.

This chapter has emphasized the need to apply a holistic supply chain approach to packaging evaluation in order to integrate and make informed trade-offs of various packaging requirements. However, to address these trade-offs, they need first to be identified and then evaluated and compared in a structured way. In the next chapter, a methodology for achieving this is presented.

Summary

This chapter emphasized the role of packaging in generating value and enabling cost efficiency by reducing product waste and improving logistics and transport efficiency. Packaging generates value in sales by attracting customers. It does so by using graphical and other design features and by providing convenience. Packaging can reduce product waste by offering protection against physical and chemical effects, temperature changes, insects and more. It can also apportion products according to stakeholder needs, thus reducing product waste in the supply chain and in the consumption phase. Product waste can also be reduced by providing consumers with information on the packaging about how to handle and consume the product. In addition, waste can also be reduced by using track & trace capabilities that provide better product control, fewer mistakes and reduced theft. Logistics and transport efficiency are affected by packaging because it, in turn, affects warehouse and transport utilization as well as production and handling efficiency.

The chapter argued for the need to apply a holistic view to the supply chain to accentuate trade-offs between various packaging requirements. This enables companies to make informed decisions. Ideally, packaging should fulfil economic, environmental and ergonomic requirements from logistics, transportation, marketing, production, sourcing and consumers; but these requirements may be in conflict. Examples were presented of common packaging system trade-offs, such as a focus on cubic utilization and the fact that packaging systems are easy to handle in logistics and production. Marketing, on the other hand, tends to focus on visibility, which is sometimes equivalent to large packaging.

The chapter discussed standardization and packaging postponement and their impacts on packaging performance. Standardization should reduce complexity, protect the interests of majorities, target selected activities, persist over time, but be subject to review and change when needed. Logistics standardization, which is affected by packaging, can be sorted into four areas: packaging (eg size and materials), transportation, material handling (storage, loading and unloading), and ICT. A low level of logistics and packaging standardization often results in greater logistics costs, but standardization can also result in lock-in effects.

Packaging postponement is one possible strategy to manage some trade-offs. The role of packaging was described in four types of postponement and speculation strategies: full postponement, manufacturing postponement, logistics postponement and full speculation. The underlying idea of postponement is to delay activities or packaging diversification until more information is available in order to increase efficiency in the earlier stages of a supply chain.

Packaging performance methodology

03

The packaging performance methodology aims to assess the overall supply chain performance of a packaging system. Based on the assessment, it is possible to determine the performance of a packaging system, of its individual packaging features, and of each packaging level. It is also possible to make well-founded decisions regarding trade-offs between the packaging features for each packaging level and to identify the improvement potential of the packaging system. The methodology applies four steps to assess the performance of packaging systems from a holistic perspective, presented in Figure 3.1.

The packaging performance methodology has been developed from the 'packaging scorecard', which was introduced by Olsmats and Dominic (2003). Since 2004, the packaging scorecard has been further developed into a complete methodology and fine-tuned in research and teaching. The methodology presented here is based on these developments and covers mapping of the packaging system and the supply chain, packaging system evaluation, and packaging system improvements, and includes tools and frameworks, shown in Figure 3.1. The packaging features to be assessed and their descriptions have also been updated to match the many roles of packaging. The scales for measuring importance and satisfaction of a packaging feature have been changed to a five-point Likert scale.

In the packaging performance methodology, the updated packaging scorecard (Steps 2 and 3) helps to collect data in a structured manner, measure the performance of a packaging system within a supply chain, evaluate and identify improvement areas of packaging systems in supply chains, and visualize underperforming packaging features both within the packaging system and between actors in the supply chain. The packaging scorecard can also be used to estimate the effects and trade-offs of potential changes in the packaging system.

Figure 3.1 Steps and supporting tools and frameworks in the packaging performance methodology

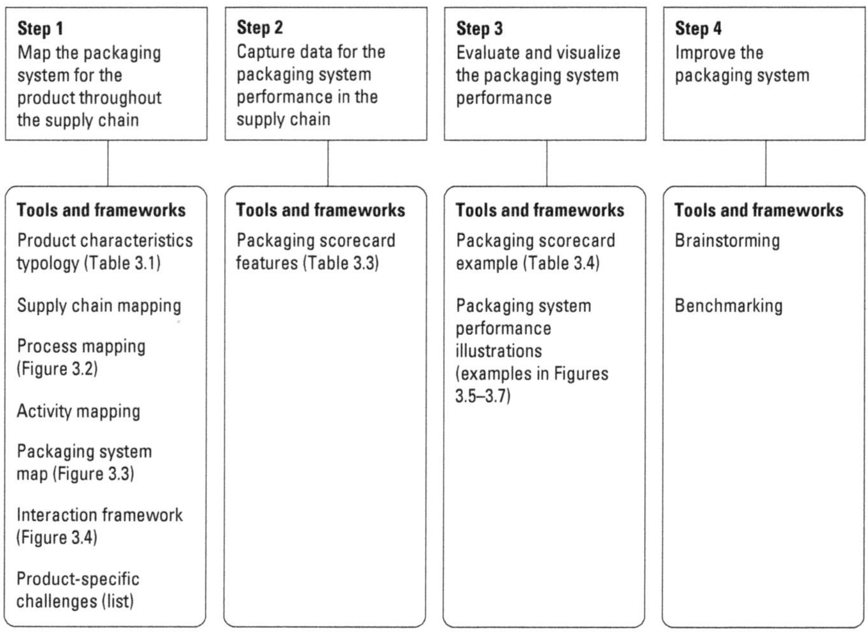

The packaging performance methodology was originally developed to evaluate and improve current packaging systems, but, with some assumptions, it can also be used for selecting packaging systems for new products. In this case, it is necessary to make assumptions about the supply chain to be used. Such assumptions can be based on an existing packed product with similar features.

The packaging performance methodology involves four steps:

1. Map the packaging system for the product throughout the supply chain.
2. Capture data for the packaging system performance in the supply chain.
3. Evaluate and visualize the packaging system performance.
4. Improve the packaging system.

Each step is presented to provide guidance on its purpose, key characteristics and challenges, useful tools and analyses, and support for practical application. Figure 3.1 presents an overview of the steps and supporting tools and frameworks in the packaging performance methodology.

Step 1: Map the packaging system for the product throughout the supply chain

The first step aims to gain understanding of the packaging system, the supply chain characteristics and the product characteristics that affect packaging. To do this, a number of tools and considerations should be used to map the product characteristics, the supply chain, the packaging system and the challenges encountered.

A Identify product characteristics' typology and their impacts on the packaging system

The product characteristics affect the extent and depth to which the packaging functions need to be assessed. For instance, the product price is critical because an expensive product may be able to absorb a higher packaging price without affecting the total price, whereas a cheap product may be heavily affected. However, if an expensive package saves more than it costs – in terms of damaged products, increased sales or improved handling efficiency – it may still be a better choice than an inexpensive one. A fragile product needs more protection than a robust one. A temperature-sensitive product, such as ice cream, needs packaging that can handle the temperature (ie packaging that is not adversely affected by cold or heat). The characteristics of the brand should also be considered, because the brand communication of the packaging should match the product, both for premium and inexpensive products. These are just examples of product features that must be described; additional ones can be found in Table 3.1. All product features and their impact on the packaging system should be described before proceeding.

Table 3.1 Product characteristics typology

Product feature	Impact on packaging (examples)
Price	Cost of packaging, labelling, packaging protection
Fragility	Packaging protection, handling
Temperature sensitivity	Packaging material
Brand	Communication
Product material	Type of packaging, packaging material
Content (fluid, food, dangerous goods, etc)	Containment
Consumer or industrial product	Communication, convenience

(*continued*)

Table 3.1 *(Continued)*

Product feature	Impact on packaging (examples)
Weight	Unitization, apportionment
Volume	Unitization, apportionment
Consumption patterns	Apportionment
Consumption volume	Apportionment
Product-specific characteristics	Depend on the specific characteristic

B Map the supply chain

After mapping the product features and their impact on the packaging system, the supply chains for the packed product should be mapped. This includes the actors and activities in the supply chain. It is preferable to use a top-down approach to this mapping, starting by defining the overall system (supply chain) and then adding more details. The actors typically include:

- production plant;
- transport;
- warehouse;
- distribution centre;
- retail store;
- consumer (not always included in the analysis).

The combination of actors is unique for each supply chain: some types of actors may be excluded in one, and more than one included in another. Each actor is involved in one or more processes in the packaging system, such as filling, labelling, picking and replenishment. All the processes and activities need to be mapped in order to understand the packaging requirements. Figure 3.2 shows the usual processes involving packaging in a supply chain. This top-level mapping exercise should be complemented with detailed activity mapping. Examples of usual activities at the producer are filling of each packaging level, manual and automatic handling, printing, sealing, labelling, palletizing, picking and loading. In the distribution centre, typical activities are unloading and loading, verification, material handling, storing, unwrapping, picking and repacking. At a retailer, common activities are unloading, verification, storing, unwrapping, picking, replenishment, material handling and waste handling. Many of these activities occur several

Figure 3.2 Usual processes involving packaging

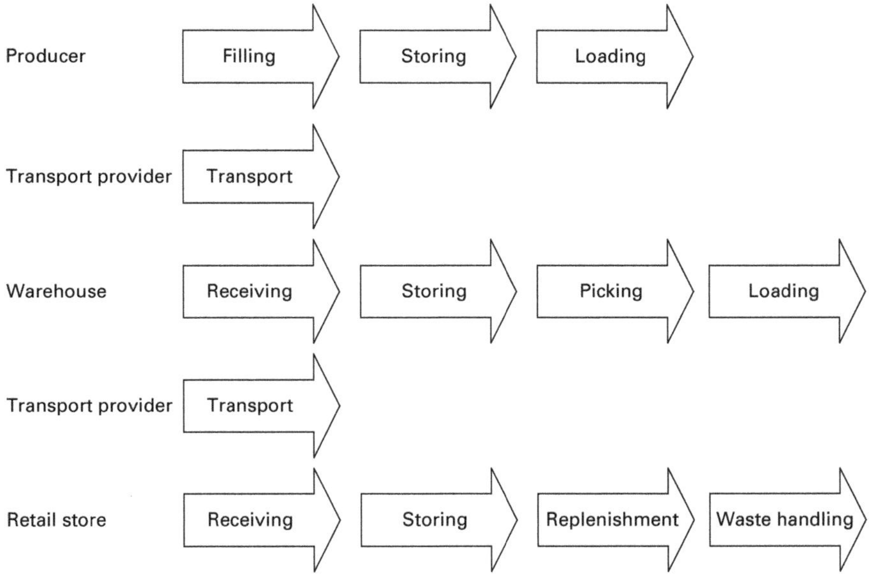

times at each actor. The following methods are recommended to collect data for the mapping exercise: the use of existing process and activity maps; interviewing staff in the different organizations; and making observations.

Keep in mind that a packed product is often involved in more than one supply chain, and many times in a network of supply chains with many variations. A product may be manufactured in one or more plants, and then sent to one or several warehouses and different stores. To manage a network of supply chains in the methodology, start from one production plant and map the supply chain network. This means top-level mapping of the diverging flows of packed products from the production plant to a number of warehouses. From each warehouse, the flows may go to several stores. The top-level mapping should then be followed by detailed mapping of one dedicated supply chain with one actor per stage. A recommendation is to map a representative supply chain in detail. However, if particularly complex supply chains exist, which top-level mapping should identify, their main characteristics should be captured as well. This helps to complement the main packaging requirements. When complete, the maps of each actor can be discussed with the other actors involved in the parallel processes. For instance, after mapping the activities in the processes performed by one warehouse, the results should be discussed with the staff at another warehouse, and differences noted.

C Describe the packaging system throughout the supply chain

The packaging system is often modified in the supply chain, for instance, when a full pallet is picked into a mixed pallet or when only primary packages are replenished in a store and the secondary and tertiary packaging become waste. To include these modifications, the packaging system should be described in each part of the supply chain, illustrated in Figure 3.3. In addition, it is necessary to understand which level of the packaging system interacts with different logistics processes. To map these interactions, the framework in Figure 3.4 is helpful.

D Map challenges for the packed product in the supply chain

Finally, product-specific challenges for the packed product need to be mapped. These can include:

Figure 3.3 Illustration of a packaging system map in the supply chain

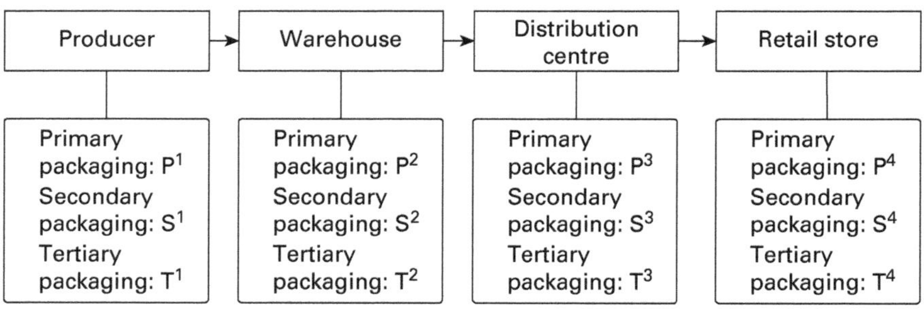

Figure 3.4 Framework of interactions between the packaging system and logistics processes in the supply chain

Actor	Process	Packaging system		
		Primary	Secondary	Tertiary
Producer	Filling	•	•	•
	Warehousing			•
Transport	Transport			•
Warehouse	Receiving			•
	Storing			•
	Picking		•	•
	Shipping			•
Transport	Transport			•
Retail store	Receiving		•	•
	Replenishment	•	•	•
	Reuse and recycle	•	•	•

SOURCE Modified from Hellström and Saghir, 2007

- theft;
- damaged products;
- damaged packaging;
- short lead-time;
- cost sensitiveness;
- high environmental impact;
- profile of the product (environmentally friendly, luxury, low cost, etc);
- weather conditions (humidity, temperature, etc);
- insects;
- biological hazards;
- waste-handling compatibility.

This completes Step 1 of the methodology. Table 3.2 summarizes the mapping tools for Step 1 in the packaging performance methodology.

Step 2: Capture data for the packaging system performance in the supply chain

The second step is about data collection using the packaging scorecard. The packaging scorecard includes 17 features in four areas, as shown in Table 3.3. The four areas cover:

Table 3.2 Mapping tools for Step 1 in the packaging performance methodology

No.	Mapping area	What to map	Tools
1	Identify product characteristics and their impacts on the packaging system	Product characteristics	Product characteristics typology (Table 3.1)
2	Map the supply chain	Actors Processes and activities	Supply chain mapping (Figure 3.2) Process mapping Activity mapping
3	Describe the packaging system throughout the supply chain	Packaging system Interactions between the packaging system and logistics processes	Packaging system map in the supply chain (Figure 3.3) Interaction framework (Figure 3.4)
4	Map challenges for the packed product in the supply chain	Product-specific challenges	The list above (theft, damaged products etc)

- packaging features with economic and environmental impacts from minimizing *product waste;*
- obtaining *logistics efficiency*;
- being *value adding*; and
- minimizing *packaging material.*

Table 3.3 Features in the packaging scorecard

Area	No.	Feature	Description
Product waste	PW1	Protection and containment	Safeguard, hold and maintain the content
	PW2	Apportionment	Manageable packaging size with sufficient product quantity
Logistics	L1	Unitization	Modularization of the packaging levels
	L2	Material handling	Enable efficient material handling
	L3	Stackability	Stacking of packages with content
	L4	Volume and weight efficiency	Utilization of the volume and weight capacity
	L5	Production efficiency	Enable efficient processing in production
	L6	Track & trace capability	Capture data and provide data to the supply chain
	L7	Reverse handling	Facilitate reverse handling, eg stack empty packaging
Value adding	V1	Product information	Provide product information
	V2	Convenience	Simplify the use of products
	V3	Promotional attributes	Promote and sell the product
Packaging material	PM1	Packaging cost	The cost of the packaging
	PM2	Packaging waste	Minimal amount of packaging waste (incl recyclability)
	PM3	Hazardous substances	Minimal amount of hazardous substances in the packaging
	PM4	Security	Protect the product from theft
	PM5	Unwrapping	Easy to open and remove used packaging

SOURCE Inspired by Olsmats and Dominic, 2003

A selection of these features should be assessed for each packaging level and for each actor in the supply chain. The packaging scorecard includes packaging features related to the six basic packaging functions (protection, containment, apportionment, unitization, communication and convenience) and a number of other features to capture the different roles of packaging in the four areas. The communication function appears in two of the packaging features: product information (in value adding), and track & trace capability (in logistics), so that both product and logistics information can be captured and communicated. Apportionment (in product waste) refers to having manageable packaging sizes with a sufficient product quantity in the whole packaging system. For example, the number of secondary packages in tertiary packaging should match the need in picking and replenishment.

A framework for selecting the packaging features to apply to the various actors in the supply chain can be developed by inserting the names of the actors in the top cells of Table 3.3, ie to the right of 'Description'. Then each packaging feature (PW1, PW2, L1 etc) that is relevant to the first actor, second actor etc can be marked in the column below each actor.

After identifying the features to evaluate for each actor, their inclusion for primary, secondary and tertiary packaging at each actor should be established. The interaction framework in Figure 3.4 shows which packaging levels to consider for each actor. By combining this with the marked packaging features for an actor, a number of packaging scorecards for each actor can be determined (Table 3.4). In this process, it should be decided whether each marked packaging feature for an actor is relevant for each packaging level in Figure 3.4. For instance, filling efficiency may be relevant for primary and secondary packaging, but not for tertiary packaging. This procedure results in a maximum of three packaging scorecards per actor. An actor that does not handle a particular packaging level (as shown in Figure 3.4) does not need a packaging scorecard for that level; for example, warehouses that usually only handle secondary and tertiary packaging.

Table 3.4 The rationale of a packaging scorecard

No.	Feature	Importance (1–5)	Satisfaction (1–5)
PW1	Protection and containment		
PW2	Apportionment		
...			
PM5	Unwrapping		

These packaging scorecards are essential to the data collection, which should assess the importance and the level of satisfaction of each packaging feature. Data collection should be carried out by conducting structured interviews with staff in the supply chain, through observations, and by reviewing logistical and packaging performance data. The latter can include damaged products and packages, handling times, fill rates and packaging waste. Based on the data collection, the researcher rates the packaging features in the packaging scorecard using the following five-point Likert scale:

Importance
1: Not important; 2: Slightly important; 3: Moderately important; 4: Important; 5: Very important

Satisfaction
1: Very low; 2: Below average; 3: Average; 4: Above average; 5: Very high

As an example, Table 3.4 presents the rationale of a packaging scorecard. The rating is subjective, which means that it is important that the same researcher or evaluator is involved in all ratings.

Step 3: Evaluate and graphically present the packaging system performance

The outcome of Step 2 is a number of completed packaging scorecards for each actor. In Step 3, these data are evaluated and presented graphically. First of all, the 'importance' values of all the packaging scorecards should be normalized (ie adjusted to a common scale) to facilitate comparisons and calculations. Normalization calculates the relative importance value for each packaging feature. If an evaluator has continuously over- or underestimated the packaging features, the relative value evens out the errors to some extent.

Normalization is done by dividing each importance value by the sum of all values in each packaging scorecard. Thereafter, it is possible to calculate the normalized average packaging performance for individual packaging levels and the packaging system as a whole for each supply chain actor:

Normalized average packaging performance

$$= \sum (\text{Normalized importance}_{PW1} * \text{Satisfaction value}_{PW1\ldots} + \text{Normalized importance}_{PW5} * \text{Satisfaction value}_{PM5}) / \text{number of packaging features}$$

PW1: Packaging feature PW1; PM5: Packaging feature PM5

To calculate the normalized average packaging performance for a packaging level (eg primary packaging), the sum of each actor's normalized average packaging performance for this packaging level is divided by the number of actors (those that interact with the packaging level). This equation is:

$$\text{Normalized average packaging performance}_{\text{Packaging level x}} = \sum \text{Normalized average packaging performance}_{\text{Packaging level x, Actor 1 to n}} / n$$

Packaging level x: primary, secondary or tertiary packaging
n: Number of actors

In a similar manner, it is possible to calculate the normalized average packaging performance for an actor. The equation is:

$$\text{Normalized average packaging performance}_{\text{Actor x}} = \sum \text{Normalized average packaging performance}_{\text{Packaging level x, Actor x}} / n$$

Actor x: Producer, warehouse etc
Packaging level x: Primary, secondary or tertiary packaging
n: Number of packaging levels

These calculations can now be used in further analyses and graphical illustrations. First, it is valuable to analyse where in the supply chain (Figure 3.5) and for which packaging levels the highest and lowest performances are. The performance of the packaging levels can be visualized in a similar figure.

Figure 3.5 Packaging performance in the supply chain

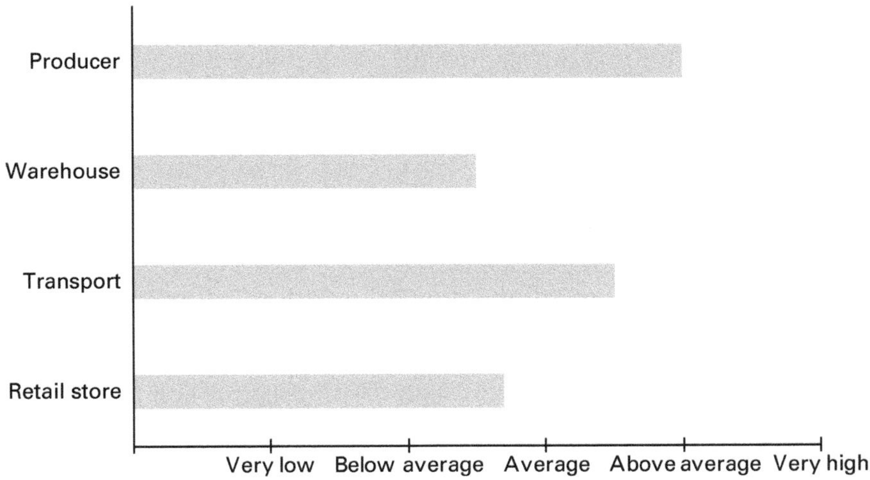

Figure 3.6 Scatter plot for primary packaging (packaging features in Table 3.3)

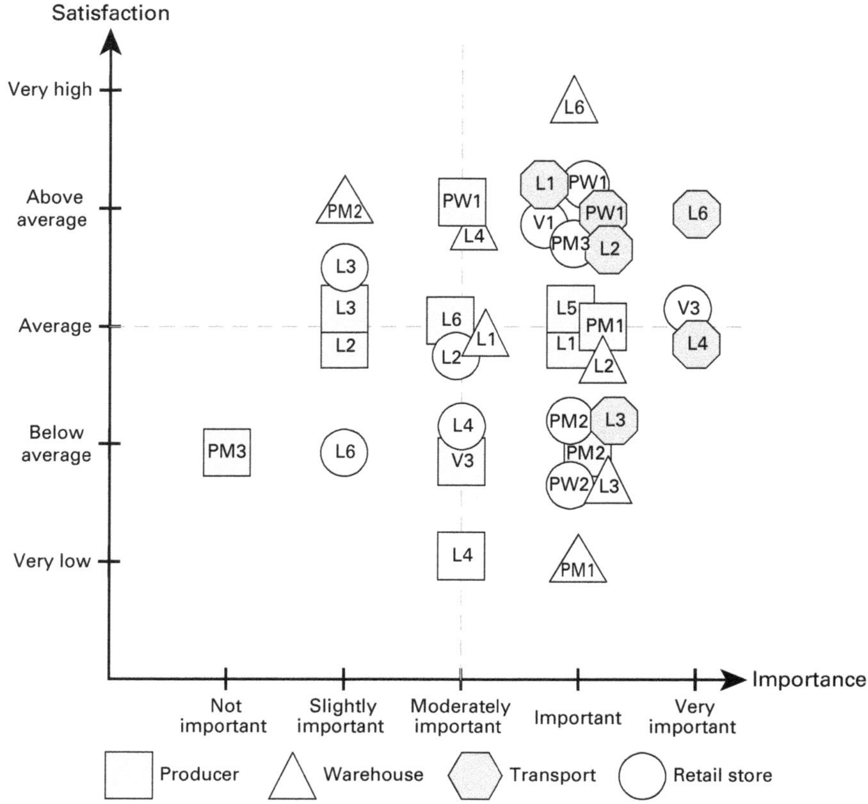

Further analyses are preferably supported by graphs. A scatter plot showing 'satisfaction vs importance' for the packaging features identified for a packaging level indicates which features need to be improved (Figure 3.6). (Explanations of the packaging feature codes are found in Table 3.3.) Most importantly, the features in the lower right square should move toward the upper right corner in Figure 3.6. Thereafter, the ones in the lower left square should be moved upwards if possible.

The structured approach of the evaluation can be summarized in five stages (I–III are illustrated in Figure 3.7):

I The whole packaging system at each actor to identify where in the supply chain the packaging system performs best and worst (Figure 3.5).

II The total performance of each level in the packaging system to identify which packaging levels the packaging system performs the highest and the lowest (a similar graph to Figure 3.5).

Figure 3.7 Illustration of Stages I–III in the evaluation

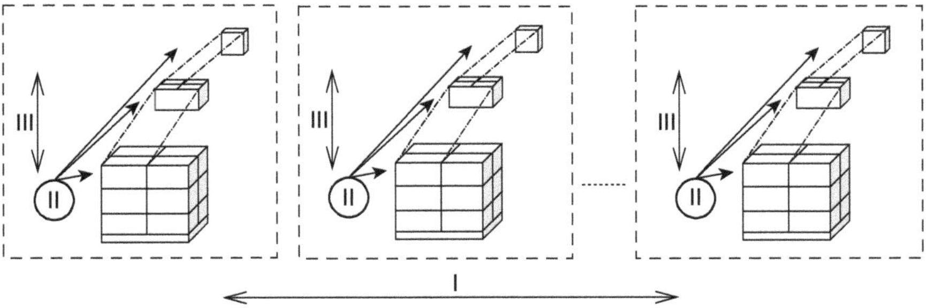

III The performance of each packaging level at each actor to identify critical packaging features in need of improvement (eg Figure 3.6).

IV Combine insights from Stages I–III to determine and summarize critical packaging features to improve at each actor and each packaging level to obtain satisfactory packaging system performance: 1) in total; 2) for each actor; 3) for the individual packaging levels. For example, it could mean that handleability and track & trace capability in the primary level, and packaging protection in the secondary level, need improvements at the manufacturer, and that stackability and volume and weight efficiency of the secondary and tertiary packaging need improvement at the warehouse.

V Apply the interaction framework (Figure 3.4) to determine potential trade-offs, which should be considered if changes are made to given packaging features.

The outcome of Step 3 in the packaging performance methodology is a list of packaging features for each packaging level and actor that should be improved to obtain the most positive effects on the total packaging system performance. For each feature, the interaction framework shows which potential trade-offs in the supply chain need to be considered.

Step 4: Improve the packaging system

The last step of the methodology results from the outcome of Step 3. Based on the list of identified packaging features for improvement, a modified or a new packaging system should be developed. To support this development, brainstorming and benchmarking exercises can be conducted.

By involving a range of organizational representatives in brainstorming sessions, different views and suggestions on modified or new packaging solutions can be found. A complementary exercise is to benchmark the packaging system primarily against the packaging system of competitors for similar products, as well as against other packed products with similar characteristics. After completing these two exercises, three to five modified or new packaging systems should be defined with consideration given to the positive packaging features and the list of improvements needed. It should be noted that a modified packaging system may include only minor changes, such as handles, or a complete new design. This depends on the current packaging performance.

After these new potential packaging systems have been developed, one of the options should be selected for comparative evaluation. This evaluation should assess the actual and physical effects of the changes. It is important to include both pros and cons, because a new packaging system can improve some packaging features and impair others. Typical assessments would include:

- To determine how many more products can be transported and handled in a more volume- and/or weight-efficient packaging system. An estimate can be made by combining this efficiency measure with the cost per product of vehicles, logistics equipment, etc. A corresponding estimate for the environmental impacts should be calculated.
- To make an assumption of the extent to which the current number of damaged products can be reduced by an improved packaging feature, such as additional protection. This should be combined with the cost of inadequate quality.
- To determine costs for changes in the amount or type of material in a new packaging system.
- To determine the effects on handling time in warehouses, loading and unloading, replenishment, etc, from changes in apportionment, unitization and stackability. This should be combined with the cost per time unit.

A complementary approach is to relate improvements or impairments to comprehensible measures. For instance, if the packaging material can be reduced by 10 per cent, it is useful to compare this effect to the packaging material's overall effect on costs and the environmental impact. Or, if a new packaging system reduces the material-handling time by one hour, then it would be interesting to know the time and cost savings per year related to the overall turnover.

The comparison of the three to five options developed can also include a packaging scorecard evaluation for each. The importance values are the same as in the original packaging system. In this final evaluation, the actual estimates support the satisfaction values for the potential new packaging systems. The packaging system with the best performance should be selected and described in detail by defining and graphically illustrating each packaging level. To make a strong case for a new packaging system, the changes must be motivated with fact-based arguments or well-motivated assumptions. The description of the new packaging system also needs to consider the whole packaging system and its trade-offs in the entire supply chain.

Organizational challenges to address for high packaging logistics performance

The packaging performance methodology provides an approach to assessing and improving the supply chain performance of packaging systems. This often leads to the identification of significant potential to reduce costs and environmental impact. However, to realize this potential several organizational challenges must be addressed.

First, the overall performance of a packaging system depends on how it functions for several actors in the supply chain. To assess this, companies need to collaborate within the supply chain in packaging development and selection processes. This means that external actors should be involved in these processes so that multiple organizations can participate in the packaging design and selection in order to capture the complexities of packaging and its overall impact on supply chain performance. From an organizational perspective, this may require a modified approach to packaging design and selection processes.

This leads to the second organizational challenge, which is to consider packaging decisions at a strategic level. If this does not happen, it becomes difficult to make organizational changes. By involving top management in packaging decisions, it is possible to embrace the strategic value of packaging.

Third, the system perspective in packaging logistics emphasizes that the focus is on the performance of the packaging system. It is essential that each of the primary, secondary and tertiary packaging units perform satisfactorily, but also that their interactions should not be forgotten. Thus, the three levels of packaging must be assessed as a system by zooming in and zooming out, as described in Chapter 1. This may be challenging for an organization or supply chain if the responsibility for different types of packaging is divided amongst different organizational units. Marketing, for instance, may be responsible for primary packaging, and logistics for secondary and tertiary packaging.

Finally, to maximize the overall packaging system performance, the interactions between the packaging system and processes in the supply chain should be managed in a resource-efficient manner. The packaging performance methodology facilitates identification and description of opportunities and obstacles within supply chains related to the current packaging system and to a modified one. However, opportunities that are highlighted at one actor often need some resources from another actor. For instance, additional handles may reduce handling time in a retail store, which requires the producer to use more costly packaging. To manage these kinds of trade-offs, it is recommended that the supply chain actors discuss cost- and benefit-sharing possibilities amongst themselves.

Summary

This chapter was about the packaging performance methodology. It helps to evaluate and improve the performance of packaging systems in supply chains, and provides effects and trade-offs of potential changes in the packaging system.

The packaging performance methodology makes it possible to determine the performance of a packaging system in a supply chain, of its individual packaging features, and of each packaging level. The methodology also makes it possible to identify strengths and weaknesses within the packaging system and between actors in the supply chain, to support well-founded decisions regarding trade-offs between the packaging features for each packaging level, and to identify the improvement potential of the packaging system. The methodology covers four steps: mapping of the packaging system throughout the supply chain; data capture; packaging system evaluation; and packaging system improvements. It provides frameworks, tools and guidelines to obtain a holistic view and make informed trade-offs in packaging systems.

The chapter presented the purpose, key characteristics and challenges, useful tools and analyses, and support for the practical application of each step. *Mapping* aims to improve the understanding of the packaging system, its supply chain characteristics, and the product characteristics that affect packaging. To do so, a number of mapping tools and considerations should be used. The *data capture* step is about data collection that uses the packaging scorecard, which includes 17 packaging features in four areas: minimize product waste; maximize logistics efficiency; add value; and minimize packaging material. This results in data capture from a number of completed packaging scorecards for each actor. In the *packaging system evaluation* step, these data are evaluated by following the structured approach presented. This results in the identification of a list of packaging system features that need to be improved. The *packaging system improvements* step uses brainstorming and benchmarking exercises to meet the improvement needs. This step results in a modified or a new packaging system.

To implement a modified or a new packaging system, several organizational challenges must be addressed. The final part of the chapter discussed these challenges.

Managing the environmental impact of packaging logistics 04

The packed product has direct and indirect effects on the environment. The direct effect arises from the packaging material, in terms of its environmental impact during production and in packaging waste. The indirect effects are related to packaging impacts on logistics and transport efficiency, and the capabilities of packaging to prevent product waste. The indirect effects are often greater than the direct ones (INCPEN, 2009), but they are often overlooked in both the literature and in practice.

In recent years, public authorities and public discussions have focused on minimizing packaging waste and use of resources. In response, legislation on global producer responsibility to minimize landfill of packaging waste and to promote recycling has been enacted. Thus, it is logical that companies tend to focus on packaging material minimization and recyclability when they aim for environmentally efficient packaging. However, in order to obtain environmental efficiencies the indirect effects should be emphasized as well.

This chapter discusses the characteristics of all three effects: packaging material, product waste, and efficiency in logistics and transport. It also presents methods to measure and obtain environmentally effective and efficient packaging systems and to minimize both the direct and indirect effects. Before these effects are investigated in detail, the chapter first discusses the roles of packaging in environmentally effective and efficient logistics and presents the relative environmental impacts of packaging material, product waste, and efficiency in logistics and transport.

Different measures can be used to capture the environmental impact. Two common ones are CO_2 emissions and energy consumption. This chapter uses the general term 'environmental impact' as much as possible, but when actual measures are necessary, it focuses on energy consumption because of data availability. This focus ensures that all energy sources are analysed, both those that generate CO_2 and those that are CO_2-free, such as nuclear and water power. It is therefore possible to identify the total energy efficiencies to be gained.

Packaging in environmentally effective and efficient logistics

In early work on environmentally effective and efficient logistics, Wu and Dunn (1995) discussed logistics decisions throughout the supply chain and their environmental impact. They highlighted the environmental impacts of logistics decisions in raw material acquisition, inbound logistics, manufacturing, outbound logistics, marketing and after-sales service. Each decision includes packaging, which illustrates the significant logistical impacts of packaging on the environment (Table 4.1).

Table 4.1 The role of packaging in environmentally responsible decision-making in logistics

Logistics decision	Environmental impact areas	Role of packaging
Raw material acquisition	Purchasing, supplier selection, and supplier location	Packaging material, packaging supplier location
Inbound logistics	Consolidation, selection of transport mode, selection of transport provider, materials handling, warehousing and backhaul management	Appropriate unitization, stackability, handleability, volume efficiency, and weight efficiency features
Outbound logistics	Same as inbound logistics, and the distribution network and packaging system selection	Appropriate unitization, stackability, handleability, volume efficiency, and weight efficiency features
Manufacturing	Inventory management and the filling process	Minimize the energy consumption in automatic handling, the same as those in outbound logistics
Marketing	Service levels and sales channels	Packaging size
After-sales service	Management of returns	Weight and volume efficiency, stackability, and unitization

Raw material acquisition involves logistics decisions about purchasing, supplier selection and supplier location. Decisions in all three of these areas influence the environmental impact of the packaging material. For instance, purchasing affects material quality, which contributes to the environmental impact of production, and in later supply chain stages, to product waste. The packaging supplier and its location affect CO_2 emissions, energy and water consumption, and other resources that are needed to produce the packaging. The packaging supplier's location also affects the mode of transportation and the distance. For instance, a global supplier may need to ship by sea and road, while a local supplier will have much shorter transportation needs.

Decisions in inbound and outbound logistics refer to consolidation, selection of transport mode, selection of transport provider, materials handling, warehousing and backhaul management. Outbound logistics also involves decisions about the distribution network and packaging system selection. In order to facilitate efficient consolidation, which contributes to reducing the environmental impact of transportation, packaging should provide appropriate unitization, stackability and handleability features. Based on the mode of transport, the packaging system design may need to be adjusted in order to minimize the environmental impact. If air freight is included, there is a greater need to focus on the weight and volume efficiency of packaging, because of high costs and environmental impact. If the mode of transport is primarily sea or rail, volume efficiency is more important, than weight efficiency as it is essential from an environmental perspective to maximize the amount of goods transported. The selection of transport provider can also affect loading and unloading routines, which consume energy. Thus, handling efficiency is a vital factor in these processes, and may include unitized packaging, and good alignment of the packaging and handling equipment. Packaging has a similar role in materials handling, where automatic handling efficiency is required for minimal energy consumption. Environmental impact decisions in warehousing and inventory also involve space utilization. A high level of space utilization results in the energy consumption from warehouse heating and lighting being divided amongst more products. Packaging facilitates high space utilization by unitization, volume efficiency and stackability. Similar packaging features contribute to efficient backhaul management.

Manufacturing decisions involve inventory management and filling processes. The environmental impacts of inventory management are similar to those in inbound and outbound logistics. The filling process should have packaging that minimizes energy consumption in automatic handling.

Logistics decisions related to marketing consider service levels and sales channels. From an environmental perspective, this affects the inventory levels and the packaging size. The smaller the packages, the greater the amount of packaging material per product. Decisions made in the sales channel determine the logistics equipment and structures with which the packaging will interact. To be environmentally efficient, these interactions should be taken into account in the packaging system design.

The final logistics decision refers to after-sales services. The environmental impact here originates in the management of returns. From a packaging perspective, it is particularly the weight and volume efficiency, stackability and unitization features that are able to make returns management efficient.

This section has demonstrated that packaging plays a significant role in delivering environmentally effective and efficient logistics. In the following sections the relative environmental impacts of these effects (logistics and transportation) are elaborated on in comparison to the direct and other indirect environmental impacts of packaging.

Environmental impact of direct and indirect packaging effects

The introduction to this chapter described how in general the relative environmental impact of indirect packaging effects is often significantly higher than that of direct effects when it comes to packaging material. This relative environmental impact varies, though, between different products and for different supply chains.

Three examples are given below to illustrate packaging material's direct environmental impact compared to the indirect effects of packaging on product waste, and logistics and transport efficiency when it comes to the energy consumption: 1) food supply chains in the UK; 2) a European supply chain for jeans from Italy to Sweden; and 3) a global supply chain of an oven from China to Sweden. The examples highlight the variations in product characteristics and supply chains. For instance, weight, volume and transport distances affect energy consumption in transportation, and the accumulated energy in the products determines how much unsold or damaged products affect the total energy consumption. Another difference is the return rate of products. Jeans, for instance, are more likely to be returned than food or ovens, which leads to additional transportation. To further strengthen the general pattern of the relative energy consumption between the direct and indirect packaging effects, the jeans and oven examples illustrate three different distribution channels. This modifies the individual share of the direct packaging effect and the different indirect effects, but the general pattern is similar.

Example 1: Food supply chains in the UK

In general, the production of food (growing, processing, production and refinement) accounts for a significant proportion of the total energy consumed in the food supply chain. To illustrate the environmental impact of packaging material compared to the indirect effects of packaging on transport efficiency and product waste for food, the energy consumption rates at various stages in supply chains are presented in Figure 4.1. This figure highlights the average share of energy consumption in different stages of supply chains for 18 categories of food products in the UK. The raw data were extracted from a report by the Industry Council for Packaging and the Environment that describes the energy consumption for one week for one person in the UK (INCPEN, 2009). It should be noted that these are indicative figures that exclude packaging disposal and waste handling.

As an indication of the relative energy consumption for food products, Figure 4.1 shows that, on average, the food supply represents approximately half of the energy consumed, while the material in the packaging system (primary, secondary, tertiary) stands for 10 per cent. Freight transport represents 3.5 per cent, and consumer passenger transport to the store is on average 1.5 per cent. This last figure is quite low because passenger transportation is shared out amongst many products (ie it is assumed that consumers purchase several products during the same shopping trip). Finally,

Figure 4.1 Energy consumption stages for food during one week for one person in the UK

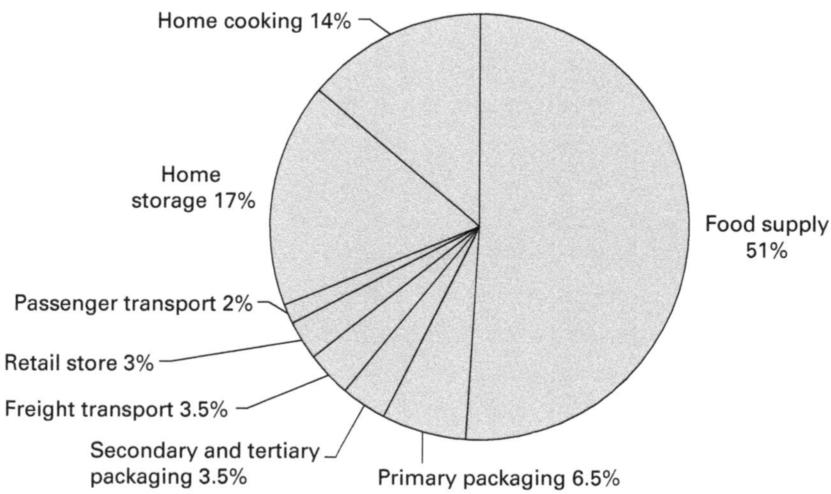

SOURCE INCPEN, 2009

home storage represents 17 per cent and home cooking 14 per cent. To some extent, packaging affects home storage activity, but not home cooking. For instance, energy usage for domestic storage can be reduced by providing sufficient ventilation or isolation for products so that they last longer.

The energy consumption in the whole system should be minimized. Sufficient packaging ensures that the accumulated energy in the packed product from growing, processing, transportation, etc is not wasted. All contributors to energy consumption should be minimized, but if prioritization is necessary, Figure 4.1 shows that it is essential to minimize food waste. It also indicates that the average energy consumption from freight transport is limited. However, note that this may vary if air freight is used and for global supply chains. Furthermore, this illustration focuses on energy consumption. If the environmental impact were measured in terms of CO_2 emissions, the results would differ because of the fossil fuel used in transportation compared to a greater share of energy sources that do not generate CO_2 emissions in the other stages. Because food supply accounts for the largest proportion of the energy consumed, it can be concluded that if additional packaging material leads to better protection and reduced product waste, this will have a positive environmental impact. If food products are damaged or wasted, the energy loss arises from the accumulated energy from the food supply, the packaging, and the stages passed in the supply chain. For instance, if a packed product is damaged in the retail store, the accumulated energy is related to the food supply, the packaging system, freight transport and parts of the retail store.

The impact of indirect packaging effects on the total energy consumption is greater than the direct impact for all food products in the data, but there are significant variations between different product categories. Figure 4.2 shows that the direct impact of the packaging material ranges from 4 per cent for bread to 46 per cent for soft drinks, but for a majority of the food categories it is between 4 and 24 per cent. The soft drinks category is the only food category where packaging material outweighs the product in terms of energy consumption. However, there is usually a high degree of recycling for soft drinks packaging, which recovers energy. Cheese, on the other hand, is an example where packaging only represents 5 per cent of the total energy, while the product itself accounts for 79 per cent. Thus, it is crucial to have sufficient packaging protection for cheese and similar products. For fresh fruit, it is important to make informed trade-offs between enough packaging material (24 per cent) and protection, since the product accounts for 56 per cent, and freight transport 10 per cent.

Figure 4.2 Share of energy consumption in various stages of food supply chains

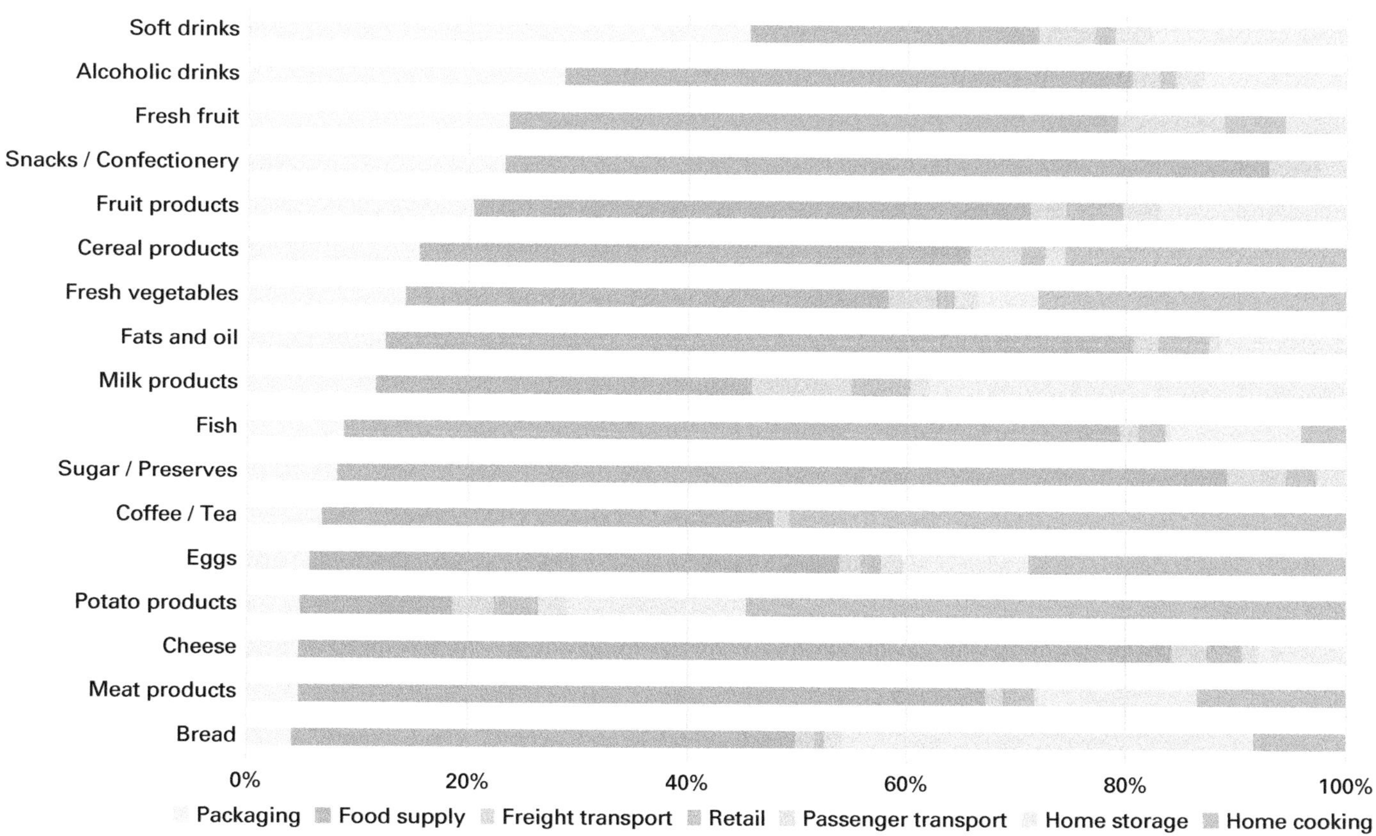

SOURCE INCPEN, 2009

As a final note, Figure 4.2 shows that the energy consumed by freight transport ranges from 1 to 10 per cent of the total energy consumption. The conclusion for food products, therefore, is that there is a clear general pattern of energy consumption in different stages, but as Figure 4.2 shows, each category of food products should be analysed in order to understand its environmental impact. As discussed earlier, the figure clearly highlights the importance of minimizing product waste. It also shows that packaging material has a significant environmental impact for several product categories. The energy consumption shares of freight transport and passenger transport are fairly low, because the products are transported together with other products. However, it should be noted that these vehicles often use fossil-based fuel, an additional reason to reduce energy consumption in transportation.

Example 2: A European supply chain for jeans from Italy to Sweden

The previous example referred to categories of food products. Other products (such as clothing and white goods) have different characteristics.

In this example, premium jeans, which can be considered as lightweight goods, are sold in a plastic bag. We assume the jeans are bought along with two other items. This average purchase assumption means that one-third of the passenger trip is assigned to the jeans. The supply chain begins in a factory in Italy, then a truck moves the products to a distribution centre in Sweden, covering a distance of 2,100 km. Then a second truck moves the jeans 10 km to another distribution centre, before a third truck transports them to the retail store. It is assumed that 10 per cent of the jeans are damaged, while product returns are negligible in the retail store.

The total energy consumption can be expressed as the sum of five factors:

- damaged products;
- packaging;
- freight transport;
- passenger transport;
- buildings.

These factors differ slightly from the food example above, because of different system boundaries. Here, the consumption stage is outside the system. Figure 4.3 shows how much each factor contributes to the total energy consumption of a pair of jeans in the supply chain from Italy to Sweden.

Figure 4.3 Five energy consumption factors for a pair of jeans in a supply chain from Italy to Sweden

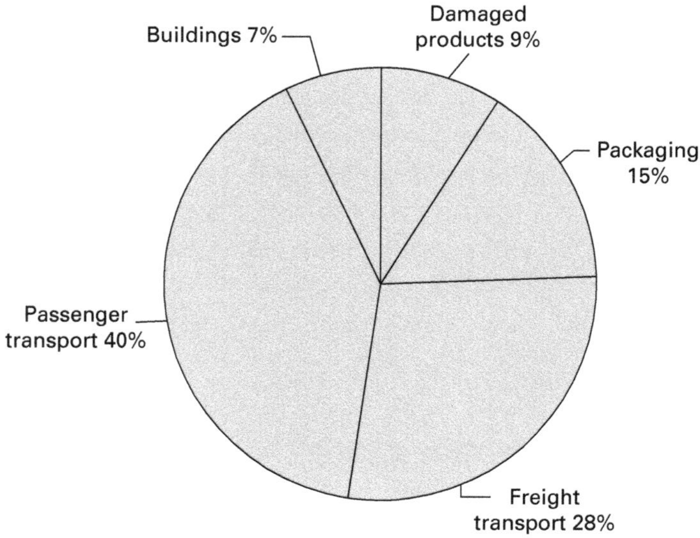

It also shows that the indirect effects of packaging are considerably higher than the direct effects for clothing as well. One difference is that the effect on transport efficiency is comparably much more important than for food, while the effect on product waste is lower. The packaging material represents 15 per cent of the total energy consumption, while the indirect effect of packaging is 43 per cent, consisting of damaged products (9 per cent), freight transport (28 per cent) and buildings (7 per cent). A considerable amount of passenger transport is done by car. If public transport, cycling or walking were used instead, the percentage goes down, and each of the other factors would increase, out of a total of 100 per cent. The relative difference between direct and indirect packaging effects would be unchanged.

Example 3: A global supply chain for an oven from China to Sweden

The oven has the following characteristics. It is heavy and transported in corrugated-board packaging. Ten per cent of the products are damaged but there are no product returns. The oven is manufactured in China, transported 200 km to a port by truck, and then shipped 19,000 km by boat to Rotterdam, the Netherlands. From there it is transported 1,100 km by truck to a distribution centre in Sweden, and finally another 320 km by truck to a retail store.

Figure 4.4 Five energy consumption factors for an oven in a supply chain from China to Sweden

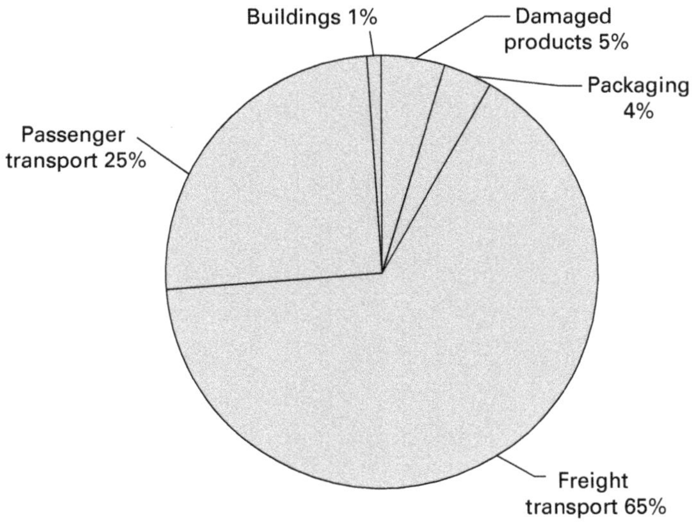

Figure 4.4 shows the energy consumption factors for the oven in a supply chain from China to Sweden. The packaging material consumes 3 per cent of the total energy. The energy consumption factors, which are indirectly affected by packaging, represent 71 per cent of the total energy. Freight transport represents 65 per cent and damaged products 5 per cent. If packaging can eliminate the damaged products, this potential reduction is greater than the entire packaging material factor. If packaging can facilitate more weight- and volume-efficient transportation, the potential in freight transport is also significant.

After this review of the total energy consumption related to packaging, the next section presents a detailed discussion of ways of managing the environmental impacts of packaging. This is divided into packaging material, effects on product waste, and packaging impacts on logistics and transport efficiency, as suggested by Molina-Besch and Pålsson (2016). Table 4.2 summarizes this structure.

Packaging material

Packaging material has a direct environmental impact arising from the energy use in its production and from packaging end-of-life, which relates to recycling, landfill and negative climate effects, such as plastics in oceans. In the EU, approximately 80 million tonnes of packaging waste is reported each year (Eurostat, 2017). The reduction of plastics in our oceans is essential because

Table 4.2 The environmental impact of packaging in three areas

Area	Packaging feature	Environmental impact
Packaging material	Balanced amount of packaging material	Minimize the amount of packaging material with consideration given to product protection and production
	Fewer mixed materials	Improved recycling efficiency
	More energy-efficient materials	Minimize the environmental impact of the packaging production process and the energy sources used in the process
	Avoidance of hazardous substances	Minimize the hazardous substances in the packaging material
Product waste	Product protection and containment	Product losses due to damages have a direct environmental impact
	Apportionment	The amount and size of packaging affect the level of obsolescence and product waste
	Packaging communication capability	Track & trace information from packaging to minimize take-backs through traceability, avoid unnecessary product waste and reduce obsolescence
	Environmental impact of consumer behaviour	Avoid product waste through easy-to-empty packaging and packaging that can be resealed
Logistics and transport efficiency	Volume efficiency	The volume efficiency of packaging affects the utilization of transport vehicles and material-handling equipment
	Weight efficiency	The weight efficiency of packaging affects the utilization of transport vehicles and material-handling equipment
	Packaging postponement with bulk shipments	Products are shipped in bulk to regional packing centres, warehouses or assembly sites
	Facilitate energy-efficient materials handling	*Stackability*: The ability to stack packages to optimize space utilization *Modularization*: To form packages into modules to facilitate unitization, stackability and volume efficiency *Unitization*: The number of packages or unit loads that are handled affects the energy consumption in materials handling *Cooling/heating needs*: Minimize energy needed to sustain the correct temperature of products by having packaging as an insulating barrier

the ocean currently absorbs about one-third of the greenhouse gases. Plastics disturb this absorption process through disruption of the ecosystem by killing birds, fish and mammals. However, 4.8–12.7 million metric tonnes of plastics are still being dumped into the oceans each year Jambeck et al. (2015).

For sustainable development purposes, it is essential to minimize waste, both from the packaging material and the products. In supply chains, particularly in global ones, the packed products are exposed to many challenges that can damage the packaging and result in product waste, such as temperature changes, moisture, humidity, insects and rough handling. In supply chains, the packaging material may be damaged or exchanged by different actors before the packed product reaches its final destination. Unless each part of the supply chain is linked to an appropriate waste-handling system, the packaging waste can end up in landfills or even in the sea. Many countries have functioning waste-handling systems for packaging waste (Molina-Besch and Pålsson, 2016). However, companies have limited insights into the whole supply chain, which results in a lack of transparency. This means that it is difficult to make an environmentally efficient packaging selection for the whole supply chain, or to control where packaging waste ends up.

Four areas should be considered when seeking to reduce the environmental impacts of packaging material production. Companies should aim to balance the amount of packaging used to provide sufficient but not too much protection. They should use as few mixed materials as possible for improved recycling efficiency. They should aim to use materials which are as energy efficient as possible. And finally, they should avoid hazardous substances.

Balanced amount of packaging material

One of the main functions of packaging is to protect the product. To achieve this, a certain amount of packaging material is necessary. However, too much results in an unnecessarily large environmental impact. When selecting packaging, the thickness of the material and the packaging dimensions should be considered. To determine the amount of packaging material to use, the combined environmental impacts of the amount of material and the number of damaged products need to be considered. An underpacked product is not sufficiently protected, which leads to considerable product waste. An overpacked product, on the other hand, may very well be sufficiently protected with less packaging material, resulting in too much material waste. To minimize the environmental impact, the amount of packaging material and the number of damaged products need to be balanced. This is illustrated in Figure 4.5.

Figure 4.5 Illustration of a balanced amount of packaging material from an environmental perspective

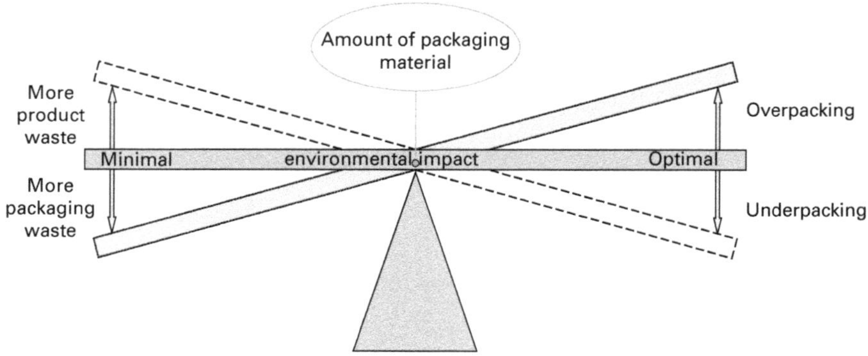

Fewer mixed materials

To facilitate efficient recycling with high-quality materials, packaging should include as few different materials as possible, as mixed materials complicate the recycling process. Unless mixed materials are necessary to perform a specific and desirable function that has positive effects on other environmental impacts of packaging, their use should be minimized. Examples of exceptions are mixed materials that reduce the total weight which, in turn, improves transport efficiency or that prolong the shelf life of the product with less product waste. For such examples, the total net effect of mixed materials should be calculated.

More energy-efficient materials

The packaging materials that are used should also be as energy efficient as possible. This means that the energy needed to produce a package is minimized in relation to the number of usages. For example, for packaging reused 25 times before recycling, the energy consumption from its production should be divided by 25 to obtain the energy expended per use; for one-way packaging, the energy expended per use is considerably higher as the entire energy consumption needs to be taken into account. As shown in Table 4.3, the cumulative energy demand (CED) from material production varies for different packaging materials. It should be noted that the table uses measurements in MJ/kg. For instance, the weight of low-density polyethylene (LDPE) used to wrap pallets or packages is low compared to the weight of a glass bottle, which means that the total energy in a glass bottle

Table 4.3 Cumulative energy demand (CED) from packaging production for common packaging materials

Material	CED (MJ/kg)
Packaging glass, white	17
Corrugated cardboard boxes	22
Packaging film, LDPE	93
EPS	90
Solid board	60
Folding boxboard	32
Kraft paper, unbleached	58

SOURCE Ecoinvent and SimaPro

may be higher than in the shrink-wrap, even though the CED is lower for glass. As an example of different packaging materials in different sales channels, the corrugated-board package for a book sold online and delivered to a pick-up point typically consumes about 4 MJ from the packaging material production, whereas the corresponding value for a plastic bag for a book purchased in store can be about 2 MJ.

Although the amount of energy required in packaging production is crucial, it is also important from an environmental perspective to be aware of the origins of the energy sources: fossil-based or renewable. In addition, environmental impacts other than energy should be considered, such as CO_2 and other emissions, water consumption and resource depletion. This information can be obtained from the packaging material suppliers.

Avoidance of hazardous substances

There are several reasons to avoid hazardous substances in packaging material. In order not to poison or expose consumers to other dangers, it is essential to avoid the migration of hazardous substances into the product, particularly in food and beverages. It is also important not to emit toxic substances from used packaging into ecosystems in the recycling or waste-handling processes. If hazardous substances cannot be avoided, their presence should be communicated through standardized statements and pictograms on labels along with safety data sheets in the package. According to the European Agency for Safety and Health at Work (2018), the most hazardous substances in packaging are: carcinogenic, mutagenic, toxic for reproduction, respiratory sensitizers, biocidal or plant protection products.

Product waste

As seen in the examples above, product waste can represent a considerable proportion of a packed product's environmental impact. For instance, for clothing sold in a plastic bag, the packaging material typically represents about 10 per cent of the total energy consumption (Pålsson *et al*, 2017). The corresponding value for furniture packed in corrugated-board boxes is less than 5 per cent. For food products, the accumulated energy in the product is often a high percentage of the total energy of the packed product (Hanssen, 1998). The difference is at least a factor of 10 greater than the energy in the packaging material. For instance, the reduction of bread waste by 2 per cent can bring about a 50 per cent increase in bread packaging material, and still result in an overall reduction of the environmental impact (Williams and Wikström, 2010).

When it comes to product waste, food presents significant problems that require specific attention. Roughly one-third of the annual global food production in tonnes is estimated to go to waste (Bond et al., 2013). In the EU, about 35 per cent of the food waste occurs from the producer to the retailer (Stenmarck *et al*, 2016). Packaging has the potential to reduce this figure significantly by providing appropriate protection. Studies indicate that in the consumption phase, which corresponds to 53 per cent, packaging can affect about 20–25 per cent of the product waste by being the right size and easy to empty (Williams *et al*, 2012). The remaining 12 per cent comes from food service.

An understanding of the integrated supply chain requirements for the packed product is required in order to minimize product waste due to out-of-date problems, moisture, temperature changes, rough handling, stackability, contamination and other factors. Four considerations should be addressed to reduce the negative environmental impacts from product protection: 1) packaging protection and containment; 2) apportionment; 3) the packaging communication capability; 4) the environmental impact of consumer behaviour.

Packaging protection and containment

To reduce negative environmental impact and minimize total energy consumption, the ability of packaging to protect and contain the product to avoid product waste is essential. This protection consists of various features because in supply chains, particularly in global ones, packed products are exposed to many challenges that may damage the packaging and result in product waste, such as temperature changes, moisture, humidity, insects and rough handling.

These challenges need to be addressed carefully. In addition, packaging can increase the shelf life of a product significantly, thus reducing product waste. Unwrapped cucumbers, for instance, are unsaleable after only three days, while 1.5 gm of plastic wrapping extends the shelf life to two weeks. In all situations with a risk of product waste, the trade-offs between packaging material and damaged products need to be evaluated, as illustrated previously in Figure 4.5.

Apportionment

Apportionment involves deciding on what are the manageable sizes of primary, secondary and tertiary packaging. From an environmental perspective, the size of the primary packaging should match the consumption patterns of consumers, in order to avoid food waste from packages that are too big. The size of the secondary and the tertiary packaging should be designed to match sales patterns to avoid out-of-date situations. Apportionment is also relevant to minimize the impacts of material handling and transportation, elements which are addressed in the next section (Packaging impacts on logistics and transport efficiency).

Packaging communication capability

Packaging has the capability to communicate track & trace data in the supply chain through labelling. Appropriate traceability information can contribute to reduced environmental impact by minimizing the amount of take-backs. As illustrated in Figure 6.2 in Chapter 6, by being able to trace all links to a contamination incident, only the products at risk need to be taken back. The packaging communication capability also helps to avoid unnecessary product waste by informing supply chain actors how to handle the packed product regarding aspects such as fragility, temperature requirements and tracking information. This last aspect supports an efficient and effective order-to-delivery process without misplaced or obsolete products. In addition, the primary packaging should inform consumers, both at the point of sale and at home, about how to store and use the product to avoid product waste.

Environmental impact of consumer behaviour

Packaging should inform and support consumers about storage and product usage conditions. This can help to prolong shelf life and avoid unnecessary product waste, and includes features that make packages easy to empty and easy to reseal in order to minimize product waste.

Packaging impacts on logistics and transport efficiency

The volume and weight efficiencies of packaging affect logistics and transport efficiency, both of which are high-emission activities, and therefore deliver high environmental impact. In OECD countries, transportation represents about 30 per cent of CO_2 emissions (ITF, 2010). Volume and weight efficiencies of packaging have been estimated to be amongst the top three supply chain initiatives to reduce these and other emissions in supply chains (Doherty and Hoyle, 2009). Packaging affects the environmental impact through improved logistics and transport efficiency through: 1) the volume and weight efficiencies of the packaging system; 2) utilization of packaging postponement with bulk shipments; and 3) enabling energy-efficient materials handling.

Volume and weight efficiencies of the packaging system

The volume efficiency of the packaging system refers to the combined cubic utilization of the primary, secondary and tertiary packaging. In other words, it refers to the amount of empty space (air) that is handled, stored or transported. This volume efficiency greatly affects logistics and transport efficiency. It affects vehicle utilization in transportation, warehouse space needed and material-handling efficiency for both filled and empty packaging.

The volume efficiency of a packaging system can be calculated by multiplying the amount of empty space in each packaging level. For instance, the volume efficiency of a packaging system with a 50 per cent volumetrically full primary packaging, 80 per cent full secondary packaging and 80 per cent tertiary packaging is 32 per cent. If a number of these packaging systems are transported in a 50 per cent full truck, the total cubic utilization of the truck is only 16 per cent. Thus, it is essential to combine the volume efficiency of the packaging system with the truck utilization to have a measure of the total potential. In practice, unfortunately, these two measurements are often handled separately. In addition, the packaging utilization is often only considered for the tertiary packaging, which means that the measurements in this case indicate 80 per cent packaging utilization and 50 per cent truck utilization.

Weight efficiency refers to the total weight of the packaging system (including the product), which affects logistics and transport efficiency. The weight efficiency always affects energy consumption in transport and material handling. It is particularly influential in situations where weight restrictions limit truckloads or the amount of goods in packaging. The latter can be because of ergonomic restrictions or weight restrictions in material-handling equipment.

By consciously striving to increase volume efficiency, great improvements in logistics and transport efficiency can usually be achieved, as illustrated by the fictional example above. Companies should consider volume efficiency in the packaging design phase, but should also measure and redesign their current packaging systems. An example of such a redesign is a tea light product at IKEA, which was first sold loosely in a plastic bag and then changed to stacked tea lights in shrink-wrapped rectangular-shaped packaging. This required a modification of the product to avoid unstable stacks and investment in new production equipment, but it increased the number of tea lights per pallet by approximately 30 per cent (Klevås, Johnsson and Jönson, 2005), which translated into 400 fewer trucks per year, more efficient storage and less handling time in the stores.

Another example is a major electronics company where a redesign of their packaging significantly reduced both its volume and weight. An illustration is presented in Figure 4.6 where the same content is packed in a much smaller box. This redesign reduced the volume by 67 per cent and the weight by 8 per cent. A third example is from a packaging company where the pallet height of the inbound deliveries could be extended from 1,200 mm to 1,300 mm, reducing transportation emissions by 14 per cent (Caraballo and Westergren, 2012). These examples illustrate that rather uncomplicated changes can have considerable positive environmental impacts.

Another example is provided in Figure 4.7 where two typical packaging solutions for tomatoes are compared. The rectangular-shaped tomato packaging is 30–40 per cent more volume efficient, and weighs 50 per cent less than cylindrical cans (size 400 gm). In this example of two real-world packaging solutions for tomatoes, 22 primary rectangular packages can fit in the secondary packaging, compared to only 14 cans. Thus, the rectangular

Figure 4.6 A redesigned package for a major electronics company with the same content

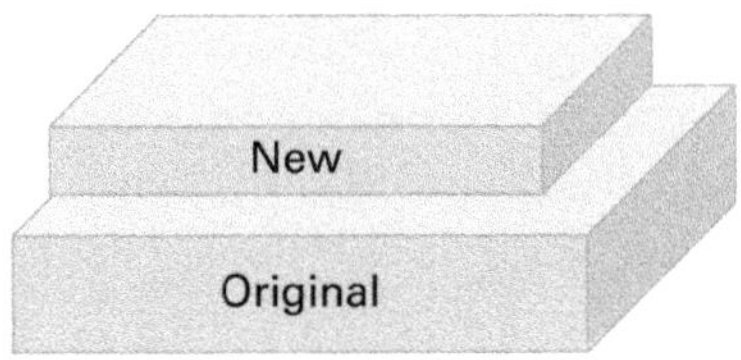

Figure 4.7 Rectangular-shaped vs cans of 400 gm tomato packaging

Length: 75 mm Width: 48 mm Height: 116 mm	Diameter: 75 mm Height: 110 mm

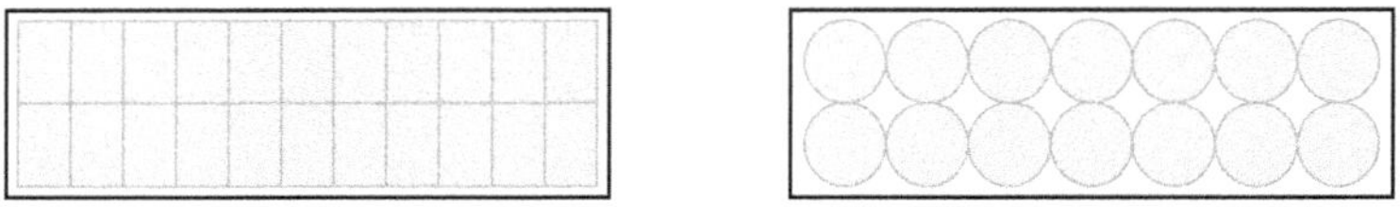

22 primary packages in the secondary packaging	14 primary packages in the secondary packaging
Surface area per package: 3.57 dm^2	Surface area per package: 3.48 dm^2

package is 36 per cent more volume efficient. However, cans are preferable in rough environments and where long shelf life is required. The best-before date for the rectangular package was one year compared to three years for the can. Furthermore, as the cylindrical shape uses less material for the same volume, the cans also have a slightly smaller surface area than the rectangular-shaped packages.

The modes of transport in the supply chain should be considered in the analysis of volume and weight efficiencies. The more energy-intensive the transport mode, the greater the effect from volume- and weight-efficient packaging. As an indication, in International Energy Agency (IEA) member countries, air freight consumes on average about 10 MJ/tonne-km, trucks 2.6 MJ/tonne-km, sea transport 0.8 and rail 0.2 MJ/tonne-km (IEA, 2017).

The packaging system can be redesigned to reduce the volume and weight of packaging for current packaging solutions by following the packaging performance methodology presented in Chapter 3. It can also be reduced by analysing overpacking (see Figure 4.5) in order to balance product protection with the amount of packaging material, and by minimizing extra items (eg using e-manuals instead of physical ones), by using lightweight packaging material as well as having larger packaging units. Another useful approach is to use bulk shipments by applying packaging postponement, which was introduced in Chapter 2 and is detailed in the next section. Regarding packaging solutions for new products, the application of concurrent product and packaging design is recommended, as outlined in Chapter 10. In this way, the packed product is developed as a system to avoid sub-optimization.

Packaging postponement with bulk shipments

As described in Chapter 2, packaging postponement can support both volume and weight efficiencies. By postponing the final packaging solution until late in the delivery process, bulk shipments with less packaging can be used in several stages of the supply chain with more space available for the products and less added weight from packaging. An example of this is beverages transported in large containers and then bottled close to the point of sale. The large containers require fewer tonne-kilometres in transport, less warehouse space and less handling than bottles. Bulk shipments can reduce the risk of obsolescence, and accordingly product waste.

Energy-efficient materials handling

Energy-efficient materials handling means having packaging that supports goods or products that are to be handled simultaneously, that avoids handling air, and that minimizes cooling or heating needs. In order to handle several products or components simultaneously or to avoid handling air, packaging systems should support stackability: the ability of packaging to stack packages to increase space utilization. Other ways to achieve this are unitization and modularization. Unitization refers to the number of packages or unit loads that can be handled simultaneously, thus affecting material-handling efficiency. Modularization is a special kind of unitization. Packaging systems are seen as modular when they can be decomposed into a number of packaging units (e.g. primary and secondary packages) that may be mixed and matched with other modularized packages in a variety of configurations to form other packaging systems. Modularization involves designing packages into modules to facilitate co-handling of different packages (eg enabling mixed pallets). Finally, packaging can have an insulating barrier to minimize the energy needed to sustain the correct temperature of products. Thus, instead of having vehicles and storage areas with heating or cooling equipment and an energy supply, the packaging system can meet this need.

Summary

This chapter presented and elaborated on the environmental effects of packaging. It emphasized the direct impacts of the packaging material on the environment during production and from packaging waste. It also highlighted the indirect effects related to the capabilities of packaging to prevent product waste, and how packaging impacts logistics and transport efficiency.

The chapter started with a discussion of the role of packaging in environmentally effective and efficient logistics. It first presented the environmental impacts of logistics decisions on raw material acquisition, inbound logistics, manufacturing, outbound logistics, marketing and after-sales service. It went on to highlight packaging's role in each logistics decision in order to illustrate the significant role that packaging plays in environmentally effective and efficient logistics.

The relative environmental impacts of packaging material, product waste, and logistics and transport efficiency vary for different product characteristics and for different supply chains. This was illustrated in three examples involving: 1) a food supply chain in the UK; 2) a European supply chain for jeans from Italy to Sweden; and 3) a global supply chain for an oven from China to Sweden.

The chapter presented a structure for addressing the environmental impacts of the packaging material, product waste, and logistics and transport efficiency, summarized in Table 4.2.

To reduce the environmental impacts of packaging material production, companies should aim to use a balanced amount of packaging material for adequate protection, but not for overprotection. They should use as few mixed materials as possible to improve recycling efficiency. They should also aim to use materials that are as energy efficient as possible and to avoid hazardous substances. To reduce product waste, companies should consider packaging protection and containment, apportionment, the packaging's communication capability, and the environmental impact of consumer behaviour. To reduce the environmental impact through improved logistics and transport efficiency, companies should consider the volume and weight efficiencies of the packaging system, utilization of packaging postponement with bulk shipments, and enabling energy-efficient materials handling.

Managing packaging waste 05

After the usage phase, packaging becomes waste. Since the packaging system consists of several parts, such as boxes, pallets, straps and shrink-wrapped plastics, which are modified throughout supply chains, the geographic location of the packaging waste varies. For instance, some parts of the packaging system end up in a warehouse, other parts in a retail store, and still others with the consumers. One general difference between packaging waste dealt with by consumers and that dealt with by companies is the way it is handled. Industrial packaging waste occurs within supply chains. Although the supply chain stages and the actors at which the location where packaging waste ends up can be described in general terms, the management and waste management practices depend on packaging characteristics that are material specific, industry specific and country specific as well as on other contextual characteristics.

This chapter describes the issues related to these challenges. It discusses packaging waste characteristics, points in supply chains for the addition of new packaging and elimination of used packaging, and takes into account the management of packaging waste for both consumer goods and industrial packaging, as well as the differences between reusable and one-way packaging. It goes on to discuss waste collection systems for packaging and management of packaging waste for supply chain actors. It should be noted that legislation and regulation are part of the industry-specific and country-specific characteristics, but these considerations are outside the scope of this book.

Packaging waste characteristics

The waste from packaging systems is related to primary, secondary and tertiary packaging. Each level of the packaging system can be exchanged several times in a supply chain. For instance, primary packaging may be put into secondary packaging for transport to a warehouse, where it is repacked into different secondary packaging. In the repacking process, other packaging parts are often replaced as well, for example, as middle layers and with shrink-wrapped plastic film. This is because the packaging system usually consists of several parts of the different packaging levels in addition to the boxes. The packaging waste that should be considered in a supply chain can typically be:

- primary packaging: packages, bags, promotional packages;
- secondary packaging: packages, lids, labels, filling material for transport;
- tertiary packaging: packages, pallets, roll containers, containers, lids, middle layers, straps, sleeves, shrink film, labels.

Each of these parts can consist of one or a combination of materials, which should be sorted and recycled according to its features.

Industrial packaging waste refers to used packaging and leftover packaging material that is considered useless in any of the businesses where packaging is handled. It originates from various processes, such as filling, assembling, manufacturing, repacking, picking, emptying and replenishment. Another type is household packaging waste, which consists of used packaging and packaging material from private homes. One main difference between packaging waste from industry and that from households is that it is easier to predict and plan the collection of packaging waste from an industrial setting than from households, as filling, repacking and other packaging-waste-generating processes can be forecasted. It is much more difficult to predict the location of consumers and which kinds of packaging materials they will throw away. For the same reason, it is easier to plan and sort industrial packaging waste than household packaging waste.

Packaging waste locations

Packaging waste occurs throughout the supply chain. Thus, each actor needs a waste collection system and organized transport of waste. Figure 5.1 illustrates packaging supplies and packaging waste points throughout the supply chain for consumer goods. The basic flowchart presented here for consumer goods sold in stores is similar for industrial packaging and supply chains for e-commerce, but differs in terms of the final destinations of the packaging waste. These differences are explained further below.

In practice, the system boundaries for studying packed consumer goods are often set from the point of filling at the producer to the consumer. Empty

Figure 5.1 Illustration of packaging waste throughout the supply chain for consumer goods

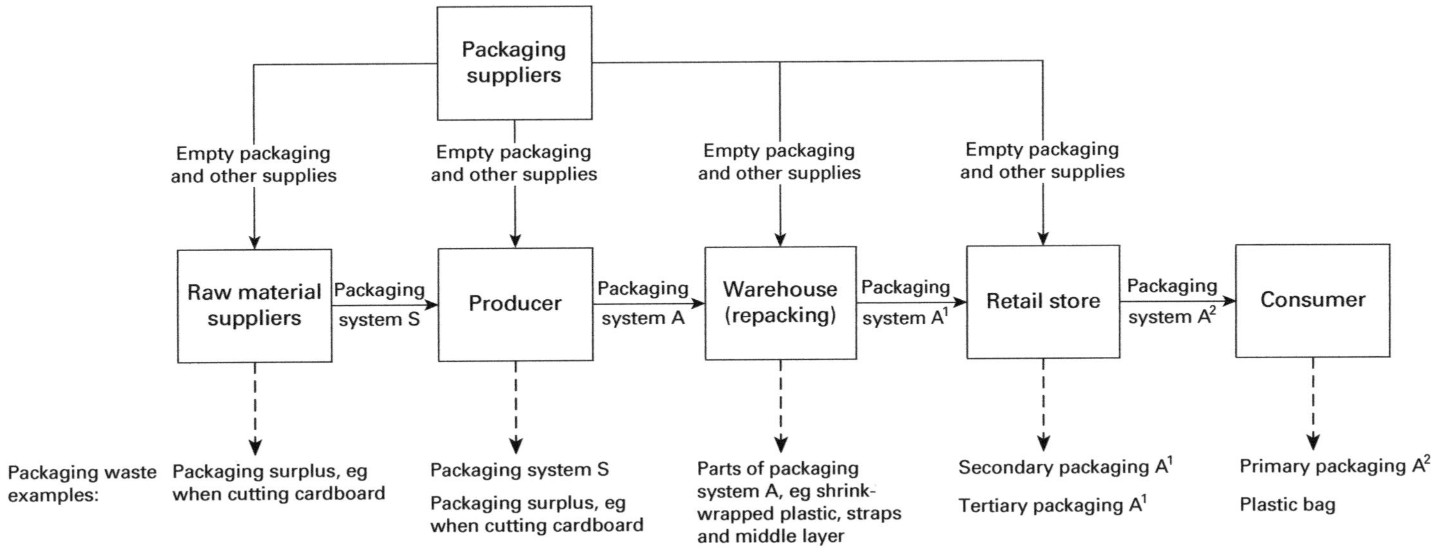

packaging from the packaging suppliers to the raw material supplies, where the packages are filled with raw material and sent to the producer, can be studied as a separate system. The waste collectors and recycling processes are outside of the boundaries in Figure 5.1.

The packaging waste in Figure 5.1 is generated as follows. The producer receives raw material from a number of raw material suppliers to manufacture the consumer goods. The raw material is delivered in packaging systems, S^1, S^2, S^3, etc. The packaging systems are emptied by the producer and then dealt with according to type: one-way packaging becomes waste while reusable packaging needs return transport to a packaging depot. Thus, the producer needs to sort and store both these kinds of used packaging. In addition, the producer receives empty packaging, packaging material and other packaging supplies, such as labels, straps, plastic wrapping, pallets and bags from packaging suppliers. After products have been manufactured, the packing process takes place. This process often generates packaging waste, examples of which include surplus from cutting corrugated board or cardboard, or from gluing, plastic wrapping and wood packaging. To handle the variety of packaging waste, several waste fractions of packaging material are necessary at a producer.

The finalized packaging system, A, is loaded and transported to a warehouse, where it is often repacked into a modified packaging system. This could be, for example, onto a new pallet with mixed products, from a pallet to a container, or from primary packaging onto a display package. In Figure 5.1, the modified packaging system is illustrated by A^1. The packaging and material removed from the packaging system become waste. This includes shrink-wrapped plastic, middle layers and straps that are not included in the modified packaging system. The modified packaging system A^1 is forwarded to the retail store.

Usually, secondary and tertiary packaging become waste in the retail store after product replenishment or after being used in the store for sale purposes. The packaging added in the store is often a bag that consumers use to carry the primary packages home. In the process of retailer replenishment or opening of the packaging system, the waste that is generated is similar to that in the warehouses, and may contain different kinds of surplus packaging materials as well as used secondary and tertiary packaging. For the products sold in display packaging, the display also becomes waste in the store. With more extensive use of display packaging, this type of waste is increasing in retail stores.

Consumers usually bring the primary packaging home as well as a bag or other package for transportation. Both of these become household waste.

Unlike industrial waste, it is less easy to predict where the packed product will be consumed or emptied. Thus, it is more difficult to determine the right waste fractions and dimensions so that packaging waste can be collected.

It should be stressed that packaging waste needs to be collected and transported from each actor in the supply chain. But it is also worth emphasizing that the supply chain illustrated in Figure 5.1 is quite a simple one. Many supply chains include several warehouses and other nodes. For industrial packaging, the consumer stage is eliminated, which means that all packaging waste ends up in a controlled industrial setting. This makes it easier to predict and correctly size the packaging waste collection and transportation systems needed.

A different supply chain setup is found in e-commerce. In delivery systems for home deliveries or collection points, the packaging waste points change. As e-commerce expands, household packaging waste increases as do the warehouse packing operations, which means that the collection systems need to be scaled up for such changes. In particular, more waste tends to end up with the consumers because of additional protective packaging, often with filling material as extra support. Other general differences are that packaging waste may increase at pick-up points and decrease at retail stores, and that more packing operations are carried out in warehouses, resulting in more packaging surplus to sort, store and collect.

Waste collection systems for packaging

Recycling systems are country specific, but the basic principles behind them all are similar. The organization of recycling in Sweden is presented as an example.

In Sweden, recycling rates are fairly high compared to other countries in Europe, Asia and the United States. Packaging waste falls under the principle of 'producer responsibility', which means that damage caused by pollution should be paid for by the actor responsible for the pollution. Thus, companies that produce, sell or import packed products are responsible for consumers' used packaging. According to FTI (www.ftiab.se), the Swedish organization that runs the national recycling scheme for packaging and newspapers, companies must ensure that:

- packaging is reusable, recyclable or able to be recovered;
- collected packaging is recycled into new raw material or energy recovered;
- customers receive the information needed for collecting used packaging;

- appropriate collection systems are easy to access and offer high service levels for consumers;
- collection data, recycling and reusable rates and material usage are reported to the Swedish Environmental Protection Agency.

From a recycling perspective, packaging is any material that 'holds, protects, handles or presents a product, or is to be used in delivering or handling a product, from the raw material to the finished product, and from the producer to the user' (Swedish Regulation 2014:1073, translated). The producer responsibility for packaging aims to limit the volume and weight of packaging and to secure recycling goals for packaging material waste.

To comply with the producer responsibility principle, Swedish industrial companies have joined together and co-founded five materials organizations and a coordinating non-profit service organization. Each organization is responsible for one type of packaging material: metal, plastics, newspapers, cardboard and glass. Each company in the system pays a packaging fee to fund their appropriate organization based on the weight of packaging material that its operations generate. The service organization is co-owned by the five materials organizations. Most of the packaging generated by the companies is covered by these organizations.

The service organization has developed a five-point programme to work toward recycling targets for each type of material set by the Swedish government. The programme defines the actors that are considered as legally responsible producers. It also includes a framework for an approved waste collection system, which should be combined with a regulatory authority. The five points in the programme refer to:

1. maintenance of recycling stations;
2. expansion of kerb-side collection for households;
3. agreements and collaboration with all municipal authorities in the country;
4. quality assessments of the waste management process with a focus on household ratings;
5. support for municipal authorities in providing sufficient information to households.

As an extra incentive, all the companies in this partnership for managing packaging waste are entitled to display the Green Dot symbol on their packaging. The Green Dot is used in many European countries with the aim of ensuring and communicating to consumers that the company fulfils its producer responsibility obligations.

Managing packaging waste

Packaging waste occurs throughout the supply chain from point of filling to point of consumption, as illustrated in Figure 5.1. To manage the packaging waste, two general principles can be applied.

First, the amount and types of waste are determined during the design and selection phases of the packaging system. To facilitate effective and efficient waste collection, these phases should follow the guidelines discussed in Chapter 4 related to packaging material. In short, this includes using as little packaging material as necessary for sufficient protection, using as few different materials as possible, and ensuring that different materials are easy to separate. It also includes using recyclable material and material with a high recycling yield (ie with a high energy-recovery factor). For in-depth discussion and details of this point, see Chapter 4.

Second, the effectiveness and efficiency of the waste collection system determine the collection rate. To collect waste from the entire supply chain, the collection system must be designed with multi-location collection points. Thus, it is essential to make sure that all waste fractions are available where packaging waste occurs. To ensure sufficient storage capacity, the size and volume of recycling containers as well as collection times and frequencies should be determined, and locations with packaging waste must have enough space for recycling containers. Space is often a scarce resource in buildings, and in facilities such as production plants, warehouses and retail stores the collection system should be planned not to disturb other operations. It should be possible to handle packaging waste from end of use to a waste bin effectively without disturbing forklift operations, picking and other material-handling operations. In production, used packaging often needs to be handled from an assembly or production line to a recycling bin. In a warehouse, there should be no packaging waste in the aisles. In a retail store, it is essential to minimize the amount of time the waste is in the store, as this can disturb sales, contaminate food and result in consumers injuring themselves. The final step in the collection process, the waste collection systems, was discussed and exemplified in the previous section.

Summary

This chapter problematized and discussed challenges related to managing packaging waste. Industrial packaging waste refers to used packaging and leftover packaging material that is considered useless in any of the businesses where packaging is handled. The waste originates from various processes, such as filling, assembling, manufacturing, repacking, picking, emptying and replenishment. Household packaging waste consists of used packaging and packaging material from private homes. It is easier to predict and plan the collection of industrial packaging waste than that of packaging from households.

Industrial packaging waste occurs in supply chains. The chapter described in general terms the supply chain stages and the actors at which the packaging waste ends up; but the management and waste management practices depend on packaging characteristics that are material specific, industry specific and country specific, as well as on other contextual characteristics. The organization of recycling in Sweden was presented as an example where packaging falls under the principle of producer responsibility. To comply, companies in Swedish industry have joined together and co-founded five materials organizations and a coordinating non-profit service organization.

The last section of the chapter presented two general principles for managing packaging waste. First, the amount and types of waste are determined in the design and selection phases of packaging systems. Second, the effectiveness and efficiency of the waste collection system determine the collection rate. The collection system must be designed to collect waste at many locations throughout the supply chain.

Managing ICT features of packaging 06

Packaging is an excellent information carrier since it usually follows the packed product from the point of filling until the point of consumption. By integrating information from labelled packaging with logistics management, the track & trace capabilities in supply chains can be improved leading to enhanced logistics performance. Decision support can be developed by capturing and communicating tracking information in real time, for example with radio frequency identification (RFID) technology. Such systems can help control material flows or predict the quality and product safety of chilled food in real time. This chapter discusses track & trace capabilities of packaging through information and communication technologies (ICT). It discusses the necessity and challenges of utilizing and managing the track & trace capabilities, the impacts on logistics performance, packaging labelling to obtain track & trace capabilities, and essential considerations in the inter-organizational implementation of these capabilities.

Track and trace capabilities enable management

Labels on packaging can provide vast amounts of detailed track & trace data. To offer an advantage, however, these data need to be utilized. After data capture, companies need to process, analyse and take action on the data. Without management and action, the data are of little value. This is exemplified by an RFID implementation in a container flow of a major paper and packaging company. It illustrates how labelling with track & trace capabilities offers supply chain visibility, but to reap the logistics benefits, it is crucial that the company utilizes and manages the track & trace data.

The paper-packaging company manufactures different types of paper products and fibre-based packaging, including paper reels that are shipped overseas in containers to a global market. Due to misplacement of containers in ports, the company had to spend considerable resources on tracking products, so decided to use track & trace capabilities through RFID technology. They placed an RFID tag on each paper reel and linked it to a unique identification code on the container. In doing so, the company gained visibility of where the paper reels were because they could verify their location. Initially, the company followed up regularly on container location, which eliminated the problem. However, over a period of time the company gradually changed routines and stopped following up the tags, which resulted once more in lost containers.

Track & trace capabilities in packaging can offer supply chain visibility, which is needed for efficiency, but this is not enough. To benefit from this visibility, companies need to take action and manage the packaging systems by utilizing the track & trace data.

The impact of ICT in packaging on logistics performance

Utilizing track & trace data can affect logistics performance positively. According to Stank *et al* (2001), logistics service performance is measured based on the abilities of organizations to:

- reduce the time between order receipt and customer delivery to as close to zero as possible;
- meet quoted or anticipated delivery dates and quantities on a consistent basis;
- respond to the needs and wants of key customers;
- provide desired quantities on a consistent basis;
- modify order size, volume or composition during logistics operations;
- accommodate delivery times for specific customers;
- match perceived logistics performance with customer expectations.

All these logistics performance measures are affected by packaging. Table 6.1 highlights the role of packaging for the various measures. For instance, by utilizing the track & trace capabilities of packaging, the time between order receipt and customer delivery can be reduced. Packed products can then be

Table 6.1 Role of packaging in logistics performance measurement

Logistics performance measure	Role of packaging
Reduce the time between order receipt and customer delivery to as close to zero as possible	Track & trace capabilities in packaging Unitization of packaging assortment Packaging postponement
Meet quoted or anticipated delivery dates and quantities on a consistent basis	Track & trace capabilities in packaging Apportionment to meet quantity demand
Respond to the needs and wants of key customers	All basic packaging functions can play essential roles depending on the types of needs and demands
Provide desired quantities on a consistent basis	A variety of packaging sizes to support apportionment to meet quantity demand
Modify order size, volume or composition during logistics operation	A variety of packaging sizes to support apportionment to meet quantity demand Track & trace capabilities in packaging to register changes
Accommodate delivery times for specific customers	Track & trace capabilities in packaging
Match perceived logistics performance with customer expectations	Track & trace capabilities in packaging Support efficient packaging handling

instantly identified and shipped. Unitization of the packaging assortment makes it easier to have mixed pallets and speeds up the palletization process, which also reduces the time taken. Packaging postponement can reduce time too, as this concept enables the final packing operation to be performed close to the market. To meet delivery dates on a consistent basis, companies need control of their goods. This is supported by track & trace capabilities in packaging. To meet quantity demands, packaging apportionment should be considered.

Logistics performance is measured by the extent to which an organization responds to the needs and wants of its key customers. Basic packaging functions have a role to play in affecting this performance. For instance, there may be a need that can be fulfilled by convenience in packaging, or to provide the desired quantities on a consistent basis, packaging can offer a variety of packaging sizes to support apportionment. The performance measures related to modification of order size, volume or composition during logistics operations can be facilitated by a variety of packaging sizes, which support apportionment. Track & trace capabilities in packaging can also be used in order to register changes in quantities. These capabilities also help to accommodate delivery times for specific customers, and to match perceived logistics performance with customer expectations. For this last logistics performance measure, packaging should support handling.

Effects of information sharing and integration on logistics performance

From Table 6.1 it can be concluded that track & trace capabilities affect logistics performance across various measures. To obtain valuable benefits from these capabilities, companies need to plan where in the supply chain the data should be captured and how to utilize them. Research has shown that companies with the highest logistics performance have several packaging identification points in the supply chain. As a result, decision-making can then be based on more timely and reliable data on movements of packed products (Johansson and Pålsson, 2009).

The highest-performing companies also share their tracking data extensively with other actors and their systems, integrating packaging labelling and registration between organizations (Pålsson and Johansson, 2009). By sharing tracking data with suppliers and customers, supply chain activities can be coordinated and integrated. Companies can utilize supply chain data by acquiring integrated systems for the registration of labels. This means that a supplier can take action on point-of-sale data, for example. However, in practice, companies currently only employ this method to a limited extent as it is quite common for companies to use different identification systems and have limited information-sharing capabilities. Figure 6.1 illustrates the extent to which Swedish manufacturing companies registered unique identities of packaging systems in a supply chain some years ago. It is more common to have control of the packaging in production than to control it upstream and downstream in the supply chain.

Effects of drivers for implementing track & trace capabilities in packaging

Most packaging utilizes track & trace capabilities to some extent. A packed product could be identified as a batch, or as a product type, and in some cases as a uniquely identified product. The latter offers the greatest potential for high logistics performance. There are many reasons for using unique

Figure 6.1 Registration structure of packaging systems in Swedish manufacturing industries: percentage of packages registered with unique identity at points in the supply chain

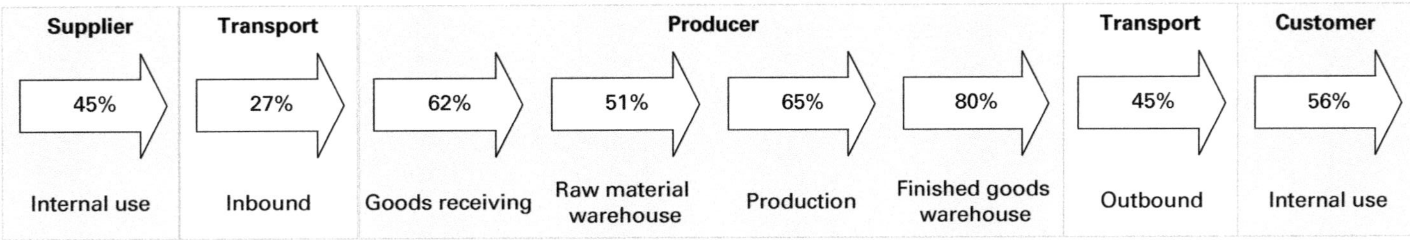

identification of packaging. In a survey of Swedish manufacturing industries, four groups of companies with different drivers were identified (Pålsson and Johansson, 2009). Companies using unique identification did so in order to: 1) coordinate material flows across organizational boundaries; 2) re-plan material flow and obtain traceability; 3) fulfil external requirements; or 4) reduce counterfeiting or theft.

These different organizational motives for implementing unique packaging identification affect supply chain integration, which in turn affects logistics performance. Companies that aim to coordinate material flows across organizational boundaries (1) and reduce counterfeiting or theft (4) are likely to achieve supply chain integration and improved logistics performance. The companies in the other two groups are less likely to increase integration and improve logistics performance. Companies that only implement these capabilities because of external requirements rarely improve their performance. Thus, if a company is mandated to implement unique packaging identification, it is advised to examine internal drivers as well.

Track and trace capabilities of packaging

Tracking is the capability to follow a packed product on its way from point of origin to point of consumption. Tracing is the capability to find the packed product between these points. Through these capabilities, packaging provides a link between the information systems and the material flow.

Keeping track of packed products is a complex matter. A warehouse can include millions of packed products that are continuously changing as they are picked, replenished and replaced. The complexity involves a combination of numerous conveyor belts, picking and storing areas, mobile and stationary material-handling equipment, etc. A similar need for track & trace occurs in other parts of the supply chain, such as transportation and production. Obviously, a track & trace system is necessary for these operations, and here, packaging labels combined with ICT can enable track & trace capabilities.

Figure 6.2 illustrates the principles of traceability for minimizing product waste in case of contamination. A well-designed traceability system can reduce the risk of a contamination incident. By analysing links between registration points in a supply chain, it can minimize the negative effects of a contamination incident, which saves both costs and environmental impacts.

In the design of a track & trace system for packaging, eight attributes should be addressed (Table 6.2), developed from Stefansson and Tilanus (2001).

Figure 6.2 Traceability of a contamination incident

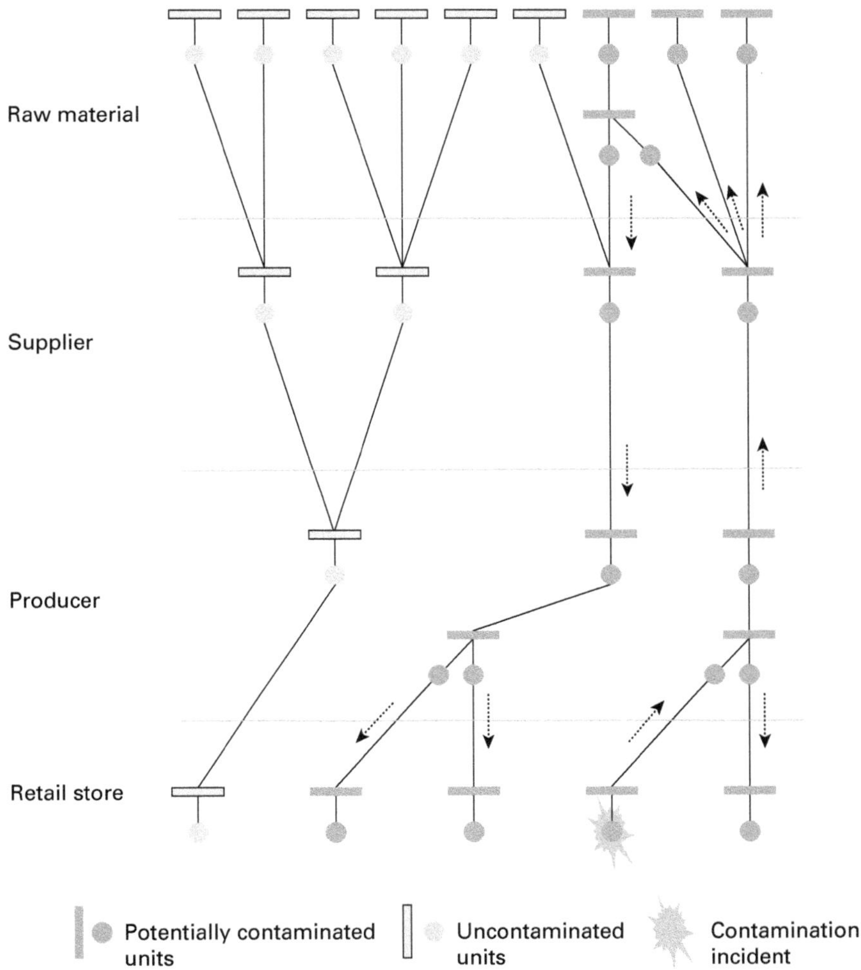

SOURCE Based on Bechini *et al*, 2008

- Based on the reasons for implementing track & trace capabilities, the supply chain characteristics and data needs, the identification technology should first be selected. The most basic identification technology uses text labels that are read manually. Barcodes are widely used as they often provide sufficient information at a reasonable price. RFID technology is also used in many applications, and provides a number of benefits compared to barcodes, such as being readable without line-of-sight, containing unique data in each label and offering multiple reads at one time, though it does incur higher investment and tag costs.

Table 6.2 Attributes to address in the design of a track & trace system for packaging

Attribute and explanation
Identification technology
Barcodes, RFID technology or text labels, for example
Scope of the track and trace systems
Defined by three dimensions: modification of place, time and form
Hierarchical level
Each discrete registration may refer to different hierarchical packaging levels
Attributes recorded
Four typical attributes to be recorded are: packaging identity, its location, changes in the characteristics of the packed product, and the point in time when the information was recorded
Time and place registration
Where and when a packed product is identified from the point of origin to the point of consumption
Activity level
Passive: Packed products are registered when they arrive or leave fixed locations *Active: The progress of a packed product is monitored from checkpoint to checkpoint and signals if something unexpected occurs*
Organization of data
Data storage can be centralized or shared by multiple stakeholders
Accessibility
Stakeholders that can track & trace packed products between the point of origin and the point of consumption

SOURCE Developed from Stefansson and Tilanus, 2001, p 205

- The scope of the track & trace system is defined by the place, time and form modifications to the packaging:
  - place modification refers to packaging being moved in transport or various handling activities;
  - time modification refers to storage that causes products to become obsolete;
  - form modification refers to conversion processes, such as production, including changes in the characteristics of the packed product, which should be specified in terms of type (eg varieties and quality attributes) and amount (eg weight, volume or number).

- The hierarchical level of the packaging system at which information should be stored must be determined. This needs to be linked to the hierarchical levels that will be registered at different locations between the point of origin and the point of consumption.
- Four attributes of the packaging system that are to be recorded should be specified: the identity of the package, its location, changes in the characteristics of the packed product (type and amount), and the point in time when the information was recorded.
- The time and place of a modification should be registered on the way from point of origin to the point of consumption.
- The design of a track & trace system for packaging needs to consider the level of activity. It can be passive, which means that packaged products are registered when they arrive or leave fixed locations. The alternative is that it is active in the sense that the progress of a packed product is monitored from checkpoint to checkpoint and signals if something unexpected occurs. An active system works well for well-defined processes in order to rapidly detect and modify errors. A passive system may be preferable in customized delivery systems with unique processes.
- Data storage can either be centralized or shared by multiple stakeholders in the cloud.
- Accessibility and security of the recorded data for various stakeholders should be addressed. Ideally, interested stakeholders should be able to track & trace packages between point of origin and point of consumption, but without violating the integrity or confidential information of other stakeholders.

Essential decision support

Packaging helps provide detailed tracking of task status, inputs and outputs. However, as described above on p. 73, logistics performance will only be improved if the track & trace data are actually used. This requires decision support for data analysis and consecutive actions that can be provided by ICT. A typology by Davenport and Short (1990) describes the organizational impacts of ICT capabilities that help provide this decision support for packaging. Table 6.3 presents examples of decision support for packaging related to eight of these ICT capabilities.

Table 6.3 Organizational impacts of eight ICT capabilities

Capability	Organizational impact	Examples of decision support for packaging
Analytical	Provide analytical methods to examine processes	Total cost analysis Fleet size of reusables Basis for planning and forecasting
Informational	Bring vast amounts of detailed information into a process	Enable just-in-time deliveries
Sequential	Enable changes in the sequence of tasks in a process, often allowing multiple tasks to be worked on simultaneously	Concurrent product and packaging production Concurrent product and packaging development
Disintermediation	Connect two parties within a process that would otherwise communicate through an intermediary	Replenishment based on point-of-sale data
Transactional	Transformation of unstructured processes into routinized transactions	Automatic ordering Facilitating co-loading of various orders
Geographical	Transferring information rapidly across large distances, making processes independent of geography	Packaging postponement
Automational	Replace or reduce human labour in a process	Tracking data in a system reduces manual tracking Potential effect of analyses of aggregated data
Knowledge management	Capture and disseminate knowledge and expertise to improve processes	Analyses of aggregated, long-term data

SOURCE Based on Davenport and Short, 1990

By considering each capability in Table 6.3, the track & trace data from packaging can be analysed and utilized in a structured way. The analytical capability of the IT system often needs to be developed to provide appropriate decision support. It could, for instance, be related to total cost analysis or life cycle assessment of packaging, calculating fleet size of reusable packaging, or making forecasts. The analytical capability should inform and support actions in the other capabilities (informational, sequential, etc). Each of the eight capabilities that supports action should be considered. In this way, the typology is a powerful tool for highlighting the potential effects of track & trace data from packaging. By addressing each capability, the full potential can be obtained in a structured way.

Implementing track and trace capabilities in packaging

The author conducted research in a six-month participant observation study of an RFID implementation with a technology provider and two major European companies: a retailer and a packaging company (Pålsson, 2006). The insights from this study led to five recommendations based on key factors for successful implementation of inter-organizational track & trace capabilities in packaging (Figure 6.3).

The first recommendation is to establish *clear motives* for each organization involved in an inter-organizational implementation project of track & trace capabilities. A motive is defined as a factor that induces a person or organization to act in a particular way. The individual motives of the organizations should then be translated into goals, which are the desired results for the implementation at both the individual organizational level, and for the agreed overall goals on the inter-organizational level. When defining motives and goals for track & trace capabilities, the organizations involved should look beyond the investment costs and instead develop a vision of the long-term opportunities for improved supply chain performance from these capabilities. Without a clear vision, it is less likely that the necessary changes in logistics processes will be made in order to obtain the potential long-term supply chain opportunities.

The second recommendation is to minimize or *reduce uncertainty* in the implementation process. Otherwise, there is an increased risk of slow project progress, unstructured project activities (eg technology tests) with ad hoc approaches, and unclear cost sharing among the organizations. Uncertainty

Figure 6.3 Five recommendations for successful implementation of inter-organizational track & trace capabilities through packaging

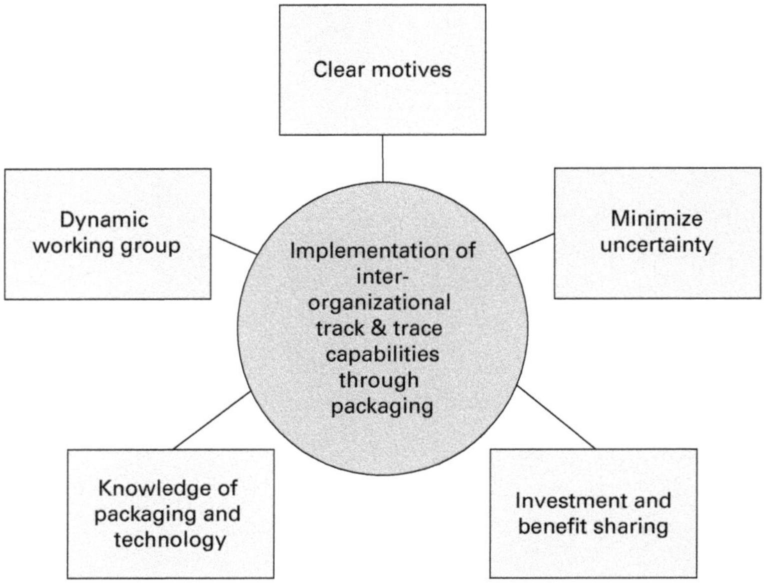

can be reduced by having top management support, a transparent information flow in the working group, and clear responsibilities assigned to each participating organization. Uncertainty is linked to the previous recommendation (clear motives), which also contributes to reduced uncertainty.

The third recommendation is to *agree on investment and benefit sharing*. The implementation of track & trace capabilities in packaging in a supply chain often provides unevenly spread benefits and costs to the organizations involved. In order to avoid this becoming a barrier to implementation, it is essential to agree on, or at least discuss, how to allocate investments and benefits early in the implementation process. However, this does not mean that the investments and benefits should necessarily be equally divided; instead, all actors should be aware of the expectations of the implementation. Without an early preliminary agreement, there is a risk that some organizations may perceive that investments, costs and benefits are unevenly balanced. This can reduce the motivation of those organizations to participate.

A fourth recommendation is to have *in-depth knowledge of packaging and technology* in the working group. This helps in the selection of appropriate technology and in understanding the potential benefits and challenges of implementing a given technology in packaging. RFID tags or barcodes, for instance, can be unpredictable in certain environments. An implementation process may also need instant access to technicians with the skills to determine the best possible placement of a label, or how the addition of labels will affect the efficiency of the packaging production process, which in turn can slow down the whole production process. Beyond technology, the need for packaging knowledge also includes its effects on other areas such as logistics and marketing. For instance, a resource with an overview of the logistical processes related to packaging in the supply chain enables the identification of all logistical opportunities associated with track & trace in packaging. The marketing requirements on packaging can also affect an implementation process. Marketing requirements, for example, can limit the possibilities for where to place labels on packaging, which affects both the automatic attachment of labels in packaging production and the ability to register labels in the supply chain.

Finally, the fifth recommendation is to establish a *dynamic working group*, because this increases the likelihood of successful implementation. As with the previous recommendation, various areas of expertise should be represented in the working group for a successful implementation process. Different organizations have different expertise, which means that they are mutually dependent. To enable an efficient implementation process, people from the different areas of expertise should be responsible and should lead their respective parts of the process. Thus, the leadership changes dynamically. Due to the complexity of the implementation process of track & trace capabilities, it is likely that unexpected challenges and possibilities will arise during the process. Thus, the working group needs to be dynamic and adaptable in order to modify conditions and continuously evaluate new scenarios.

Summary

Packaging is an excellent information carrier because it follows the packed products from point of filling to point of consumption. The utilization of data from labelled packaging in logistics management enables improved track & trace capabilities in supply chains. This can enhance logistics performance.

This chapter discussed track & trace capabilities of packaging through information and communication technologies (ICT). It considered the necessity for and challenges of utilizing and managing the track & trace capabilities, the impacts on logistics performance, packaging labelling to obtain track & trace capabilities, and essential considerations in the inter-organizational implementation of these capabilities. The usage of ICT to keep track of packed products is a complex matter, and the design of a track & trace system for packaging should address the impacts of eight capabilities (technology, scope, hierarchy, attributes, time and place, activity, data organization and accessibility).

The chapter also discussed the potential of track & trace capabilities through packaging to affect various measures of logistics performance. It emphasized that to take advantage of the track & trace data, the data need to be utilized. It showed that companies with the highest logistics performance have several packaging identification points in the supply chain so that the decision-making can then be based on more timely and reliable data about the movements of packed products. The highest-performing companies also extensively share tracking data and inter-organizationally integrate packaging labelling and registration systems. The utilization of tracking data requires decision support in ICT systems for data analysis and consecutive actions. To structure decision support for packaging, a typology of organizational impacts of ICT capabilities was presented. This includes decision support for packaging related to eight ICT capabilities (analytical, informational, sequential, disintermediation, transactional, geographical, automational and knowledge management). By considering each of these capabilities, the track & trace data from packaging can be analysed and utilized in a structured manner.

Finally, the chapter discussed how to implement track & trace capabilities in packaging, particularly in an inter-organizational context. The chapter presented five recommendations for the inter-organizational implementation of track & trace capabilities: clear motives, minimization of uncertainty, agreement on investment and benefit sharing, knowledge of packaging and technology, and dynamic working group.

Reusable packaging systems 07

Reusable packaging is designed to be reused without weakening its protective function. Thus, each package is usually stronger, heavier and more costly than one-way packaging. From a logistics perspective, reusable packaging requires reverse logistics and the management of this process.

Reusable packaging has many industrial applications and usage areas, both in terms of secondary and tertiary packaging. The automotive industry, for instance, often uses a pool system of plastic crates for their component supply; food producers and agriculture often use plastic crates for bread, fish, fruit and vegetables, both in internal material handling and in distribution. Figure 7.1 shows a material-handling example at a farm. Beverages are also frequently distributed in reusable packaging. Pallets, such as Europallets, are a common tertiary reusable packaging type.

Characteristics of reusable packaging systems

A comparison with one-way packaging illustrates the characteristics of reusable packaging systems:

- Reusable packages have a higher investment cost compared to one-way packages because they need to be more robust and have better protective features. However, this cost should be divided among multiple usages as the operational costs of reusable packaging involve more activities than one-way packaging.
- Reusable packaging incurs operational costs for handling, storage, cleaning, repairing and management, whereas one-way packaging typically only incurs handling and storage costs.
- Reusable packaging requires more management input for reverse logistics and the avoidance of package theft, damage and misplacement. These management issues are discussed at the end of this section.

Figure 7.1 Reusable packaging in material handling at a farm in South Africa

From an environmental perspective, reusable packaging can reduce packaging waste in the supply chain. However, to determine whether the environmental impact is better or worse than one-way packaging, a systems approach is needed which includes other environmental impacts such as transport efficiency and effects on product waste. In the current literature, many authors do not take a systems approach. Instead, they mistakenly assume that reusable packaging is more environmentally efficient than one-way packaging simply because it is used several times. However, in some contexts, reusable packaging is more environmentally beneficial and in others, one-way packaging is better.

To compare the environmental impact of reusable vs one-way packaging systems, the volume efficiency of the packaging system as well as packaging material, material-handling, transport and waste-handling factors should be analysed. For insights into how to analyse these factors and compare reusable to one-way packaging, see Chapters 8 and 9. From an ergonomic perspective, reusable packages are often more robust with better stability, but are also heavier than one-way packaging. They are also more likely to use standardized dimensions, which can provide co-loading and other efficiency benefits, but also risk a lock-in effect, where it becomes difficult to modify the packaging to cater for new types of products, new customers and changes in order sizes.

When reusable packaging is suitable

The suitability of reusable packaging depends on supply chain characteristics. Reusable packaging uses a closed-loop process, which includes return transport. This is not necessary for one-way packaging, which does however generate waste requiring transportation. To minimize the costs and environmental impact of the return transport, the distance should be as short as possible, meaning that reusable packaging systems are more competitive on short shipping distances. The cycle time (the total time from the point of filling until the reusable package is ready to be used again) also affects the costs, and should be as short as possible in order to minimize fleet size. The fleet size can also be affected by good labelling and a track & trace system, which can eliminate lost and misplaced packages as well as optimize the cycle time through control management. Further, for reusable packaging to be economically and environmentally efficient, the handling, sorting and cleaning operations of empty packages also need to be efficient. Finally, a strong supply chain leader and an industry consortium to standardize reusable packaging will both provide support for reusable packaging systems (Twede and Clarke, 2004).

The suitability of reusable packaging also depends on specific contextual variables which affect its cost and environmental efficiency for industrial use: volume efficiency, material flow type and packaging size.

The *volume efficiency* of the packages greatly influences the cost and environmental efficiency of reusable packaging, as the characteristics of the packaging content (weight, size, shape) determine the extent to which the inner volume of a package can be utilized. For instance, the inner volume of one-way packaging with perpendicular sides is greater than that of reusable packaging with conically-shaped sides. The smaller the packages, the greater the effect. Hence, small conically-shaped packages are more likely to be more cost and environmentally efficient in one-way than in reusable packaging, as more products or components can be stored, handled and transported simultaneously in the one-way package.

The *material flow type* has a similar effect on cost efficiency. The material flow type refers to the handling and movement of packaging in manufacturing plants, which is also affected by the number of products or components in a package with the same outer dimensions. Components with more internal handling (eg repacking, picking and movements of packed products) are more cost efficient in one-way packaging from a material-handling perspective, if the inner volume is greater. This is because the volume efficiency of one-way packaging reduces the number of packages that need to be handled, thus reducing the handling time per component. At the same time, larger packages can lead to reduced inventory turnover, which increases the cost of capital (as explained in Chapter 8), but it usually does not outweigh the reduced handling cost.

Finally, the *packaging size* affects the cost and CO_2 efficiency of both reusable and one-way packaging. Smaller packages are less likely to be cost and CO_2 efficient in conically-shaped reusable packaging. There are several reasons for this, but the greatest impact usually comes from volume efficiency of load carriers. This means that if reusable packaging is conically shaped, there is more space between packages on a load carrier for smaller packages than for bigger ones.

Management of reusable packages: an important issue

Reusable packaging requires good management of reverse logistics flows. Otherwise, there is a risk of long cycle times, misplacement, theft and damaged packages, which leads to high costs and supply uncertainty. This, in turn, results in a need for a bigger fleet of reusable packages, which is costly in itself and requires more storage space and washing capacity. Thus, appropriate management capabilities to track & trace reusable packaging at any location in the supply chain can contribute significantly to the efficiency of a reusable packaging system.

Types of reusable containers

Reusable secondary packaging, often referred to as reusable containers or trays, can be divided into four basic types: stackable, nestable with lids, stackable and nestable, and foldable. These reusable container types have different characteristics that are suitable for various contexts. Table 7.1 compares the characteristics of the four basic container types. All reusable containers are usually heavier than one-way packaging of similar size, but are also stronger and sturdier. The four basic types are described as follows:

- Stackable containers are often the strongest and sturdiest, and can be stacked without lids. These containers provide good volume utilization in deliveries, as the inner volume represents a large share of the outer container volume. Thus, fill rate in stackable containers is high. Unfortunately, stackable containers cannot be nested, which leads to low volume efficiency in return transport.
- Nestable containers need to have lids in order to be stackable. The volume efficiency in delivery transport is slightly lower than for stackable containers, as the sides of nestable containers have a conical shape and the lid needs some space. On the other hand, the volume efficiency in return transport is much higher due to the nesting capability.

Table 7.1 Comparisons of basic characteristics of four types of reusable containers

Characteristic	Stackable	Nestable with lids	Stackable and nestable	Foldable
Strength	Strong and sturdy	Sturdy	Sturdy	Fairly sturdy
Stackability	Yes	Yes (due to the lid)	Yes	Yes
Volume efficiency in delivery transport	High	OK	Fairly OK	High
Inner vs outer volume	High volume utilization	The nesting possibility reduces the inner volume The lid needs space	The nesting possibility reduces the inner volume	High volume utilization
Volume efficiency of empty packages	Low	OK	OK	High
Special requirement	—	Conical shape of boxes requires lid for stacking	Orientation determines stacking or nesting	Require erection and folding

- Stackable and nestable containers can be used in two directions. The orientation of the container determines whether it is stacked on or nested in another container. The nesting capability reduces the inner volume to some extent, but it is still quite volume efficient both in delivery and return transports.
- Foldable containers are typically the least sturdy, but are still often strong enough. Foldable containers are stackable and require minimal space in return transport, because of the folding capability. The volume efficiency in delivery transport is high since the sides can be perpendicular to the bottom, which maximizes the ratio between inner and outer packaging volume. One potential disadvantage of foldable containers is that they need to be erected and folded, which often requires manual labour.

All four container types can have additional features. Nestable containers require lids to be stackable, but lids can also be used for other types of containers, to protect products from dirt, for example. Another feature is to include handles, which can be designed in a number of different ways. One specific type of handle is a bail arm, used both as a handle and to improve the stackability of a container. Reusable containers can also be designed with perforation, which can have multiple purposes: it can facilitate airflow to sustain product quality by maintaining the same temperature and level of freshness for all products; and can provide visibility of the content and filling levels, and reduce the container weight. When it comes to the inclusion of labelling, as described in Chapter 5, reusable containers are better able to include more costly identification technologies (eg RFID) as the containers are used multiple times. The labelling can be in-moulds, label pockets or designed with windows. Another feature is slide stacking, which means that a container can slide into its stacking position on the container below. Reusable containers can also be designed so that dual heights can be utilized. In transportation, for instance, the open top is headed up vertically, but on a shelf, the open side can be in a horizontal direction to support picking. Dividers can be used to separate sections in the containers to further support picking.

Control methods for reusable packaging systems

Reusable transport packaging is part of reverse logistics, which is defined as 'the role of logistics in product returns, source reduction, recycling, materials substitution, reuse of materials, waste disposal and refurbishing, repair, and remanufacturing' (Stock, 1998, p 2). Thus, reverse logistics includes both materials management and physical distribution where reusable packaging plays a role. A reverse logistics system for packaging involves transportation, cleaning, maintenance, storage and administration of packaging. In the reverse logistics system, packages are moved from a sender (package filler) to a receiver (customer), and then to another sender (which could be the same as the previous sender). Between the sender and receiver and the next sender there can be nodes in terms of warehouses and washing facilities. This is illustrated in Figure 7.2.

The reverse logistics system for reusable packaging can be organized in three ways (Kroon and Vrijens, 1995). These are: 1) switch pool systems; 2) systems with reverse logistics; and 3) systems without reverse logistics. The characteristics of these ways of organizing the reverse logistics system for the reusable packaging are summarized in Table 7.2.

Figure 7.2 Senders, receivers and possible nodes in a reverse logistics system for reusable packaging

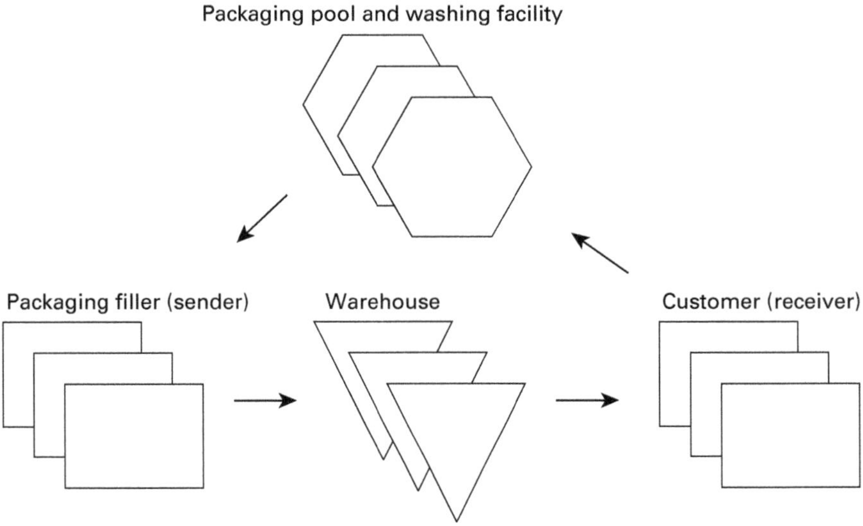

In the switch pool system, every partner has an allotment of packages. This means that each partner is responsible for the quality of its own allotment, including quality control, cleaning, maintenance and storage. The partners in switch pools are senders and receivers, and sometimes transport providers. In the latter case, the transport provider also has an allotment of reusable packaging. This simplifies the management and administration for the sender because the transport provider replaces packages with empty ones each time a load is picked up. In the case of only a sender and a receiver, transport providers transport filled packages from the sender to the receiver and empty ones in the opposite direction. However, it is not always the same number of packages going in both directions each time. There must be an agreement, though, that all empty packages are returned to the sender in the long term.

In a switch pool system and in a system without reverse logistics (described next), packaging can be dedicated to each sender or shared between different senders. In a dedicated setup, senders have their own type of packages (which can be similar, but can also vary), whereas a shared setup means that a common packaging fleet is shared. The reverse logistics system always uses a shared setup. By comparison, a dedicated setup is easier to manage as it involves fewer organizations using the packaging. This means that the shared setup requires more advanced management and IT systems. On the other hand, the total safety stock is higher with dedicated packaging, as

Table 7.2 Characteristics of three types of systems for reverse logistics

	Switch pool			With reverse logistics		Without reverse logistics
Basic principle	Every partner has an allotment			Reverse logistics by agency		Rental of the containers
Responsibility	Every partner is responsible for his own allotment			Agency		Sender, also for the reverse logistics
Investment	Shared among users			Agency		Agency
Partners	Sender, receiver	Sender, transport provider, receiver	Agency, sender, transport provider, receiver			Agency, sender
Control system	Direct switch	Exchange-per-exchange switch	Transfer system	Depot system with booking	Depot system with deposit	Rental of the containers
Benefits	Simple administration	Simple administration Control of packages in the system	The sender has good control of own packages	Minimal administration for packaging users Economies of scale for cleaning and maintenance Cost-efficient return transport	Cost-efficient return transport Track & trace system is not needed Short cycle time Deposit covers lost packages	No investment for users Low economic risk for users Easy-to-handle flexible packaging volumes
Drawbacks	Risk of long cycle times Risk of lost packages	Extra packaging inventory at the transport provider	Internal resources for administration for the sender No economies of scale for cleaning and maintenance	Advanced IT structure Frequent reporting	Advanced IT structure Frequent monetary transactions	Rental fee to the agency No economies of scale for cleaning and maintenance Senders need own cleaning and maintenance facilities

SOURCE Developed from Lützebauer, 1993 and Kroon and Vrijens, 1995

Figure 7.3 Principal transportation flows in a shared set-up of reverse logistics

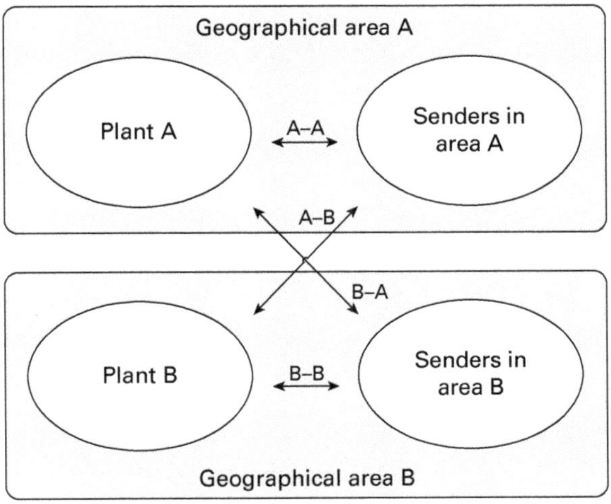

senders need to hold their own stock. In a shared setup, the safety stock can be held centrally.

Based on a model in the automotive industry, Zhang *et al* (2015) calculated and compared the cost of dedicated and shared setups. They found that the shared setup is generally more cost efficient due to lower transportation volume, pipeline inventory and safety inventory. To maximize the cost efficiency of a shared setup, all senders should use the same type of packages as much as possible. In addition, as illustrated in Figure 7.3, the transportation volume both within (A–A and B–B) and between (A–B and B–A) different geographical areas should be balanced as much as possible for each package type.

In the reverse logistics system, an agency owns and manages the reverse flow of reusable packaging. The other partners in the reverse system are senders, transport providers and receivers. The receivers should store empty packaging until the agency determines that a cost-efficient amount can be collected and returned. Reverse logistics can use one of three different control systems:

- The first option is a transfer system in which the sender has a key role. Each sender has control over the number of reusable packages, as it uses the same containers repeatedly. The sender accounts for cleaning, maintenance, storage, administration and keeping track of its packages.
- A second method is a depot system with booking. This is characterized by all available empty packages being stored in packaging depots

where they are cleaned and maintained. Senders order reusable packages from the depot, and after being used and transported to a receiver, the empty packages are returned to the packaging depot. The role of the central agency is to keep track of and control the packaging flow in a booking system. All senders and receivers have accounts in the booking system. For each transaction, the number of packages is registered in the account. For instance, when a sender gets a number of packages, the account debits this quantity for the particular type of packaging. When these packages or another quantity are sent to a receiver, the credit of the actual quantity is drawn from the sender's account and debited in the receiver's account. Each transaction includes type of package, quantity, and name and address of the organization.

- The third option is a depot system with deposit. The basis of this system is similar to the second method in that available empty packages are stored in packaging depots. The difference lies in the packaging ordering system. In the depot system with booking, packaging is booked, while the third option uses a deposit system. Each organization pays an administrative fee to the agency. This method has three benefits. The deposits provide incentives to return the packages, which means that it is unnecessary to have a track & trace system for this purpose (other purposes may still benefit from a track & trace system) and the cycle time of packages in use can be shortened. The deposit also covers lost packages. In this alternative, a deposit for each package is set, which should be equal to or higher than the packaging value (to cover lost packages and support returns). Each time a packaging transaction takes place, a deposit is paid. Figure 7.4 illustrates the flow of packaging and the opposite direction of deposits. The idea is that the sender pays the agency, the receiver pays the sender, and the agency refunds the deposit to the receiver.

Finally, in the system without reverse logistics, an agency owns a pool of packages. Here, the senders do not have to invest in reusable packaging, which reduces their fixed costs. In this system, the packaging fillers (senders) pay a rental fee to the agency for the number of packages needed. During rental, the sender has full responsibility for cleaning, maintenance, storage, and keeping track of the packages internally and in reverse flows from customers (receivers). When all or a given number of packages are surplus, the sender returns them to the agency.

For all three types of reverse logistics systems, it is necessary to determine the fleet size of the packages in the system. To calculate the fleet size, consideration should be given to cycle times, number of customers, variation in

Figure 7.4 A depot system with deposit

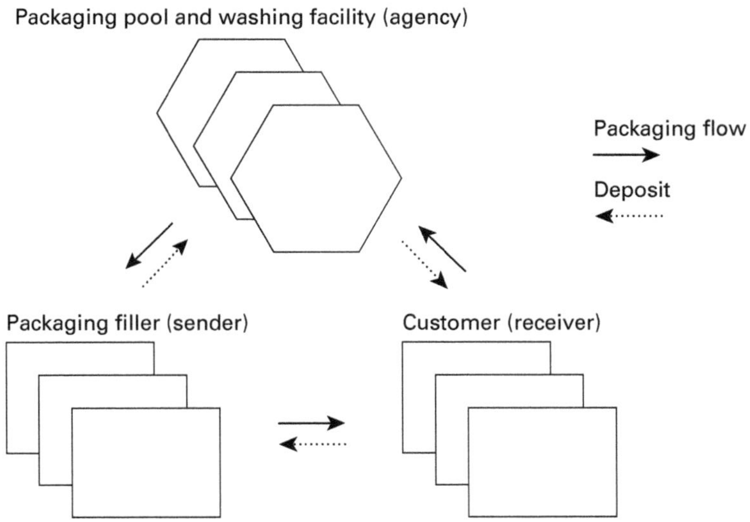

demand, share of lost packages, time for maintenance, and safety stock. It is also necessary to determine the number and location of depots. To do so, calculations for depots should consider the number and location of senders and receivers, and the number of packages at these partners. The calculation should also consider investment and operational costs of a packaging depot, particularly in terms of economies of scale. In addition, even though it is the control system of each type of reverse logistics system that determines the general principles for distributing, collecting and relocating packaging, the details behind these principles need to be agreed upon. For instance, in the switch pool system, bundled empty packaging is returned to the sender, but in order to avoid cycle times that are too long and thus a large fleet of packages, the minimal frequency of return transport can be agreed upon. Finally, fees for service, distribution and collection need to be determined for all the systems. This should be based on investments, operational costs and potential profits.

Summary

Reusable packaging has many industrial applications and usage areas, in terms of both secondary and tertiary packaging. This chapter first discussed the characteristics of reusable packaging systems compared to one-way packaging systems. This showed that reusable packaging needs to consider operational costs from handling, storage, cleaning, repairing and management, whereas one-way packaging typically needs only to consider handling and storage costs. In particular, reusable packaging requires more management of reverse logistics to avoid theft, damage and misplacement of packages. The chapter also clarified that in order to determine whether the environmental impact of reusable packaging is better or worse than one-way packaging, a systems approach is needed.

To determine whether a reusable packaging system is more or less costly, or has comparatively more or less environmental impact than a one-way packaging system, contextual variables and supply chain characteristics were discussed. The three contextual variables are volume efficiency, material flow type and packaging size. The supply chain characteristics are related to transport distance, cycle time, efficient operations of empty packages, existence of a strong supply chain leader to manage reusable packaging, and an industry consortium that can manage the standardization of reusable packaging.

The chapter went on to describe four basic types of reusable packaging: stackable, nestable with lids, stackable and nestable, and foldable containers. It also discussed additional features for these basic packaging types (eg lids, handles and perforation). To manage these reusable packaging types, a reverse logistics system is needed.

In the final part of the chapter, the characteristics and organization of three types of systems for reverse logistics were discussed. In a switch pool system, every partner has an allotment of packages. The system with reverse logistics is characterized by an agency that owns and manages the reverse flow of reusable packaging. In the system without reverse logistics, the basic principle is that the users rent reusables from an agency that owns a pool of packages.

Industrial packaging evaluation

08

Industrial packaging is used in the supply of material. The main purposes of industrial packaging are to protect the products and facilitate efficient production, material handling, warehousing and transportation processes. Its main differences from consumer packaging are the lack of promotional and sales features. The features of industrial packaging have an impact on the economic, environmental and ergonomic performance of supply chains. The impact on the ergonomic performance is mainly seen in material handling, such as in moving and emptying packages, picking and repacking. However, whilst the ergonomic aspects of industrial packaging are important, this book focuses on the economic and environmental factors. For the purposes of this chapter, therefore, it has been assumed that the ergonomic requirements for the packaging system have already been fulfilled, and that the economic and environmental performance can now be evaluated.

For a complete environmental analysis of packaging's impact, several measures are needed, such as CO_2 emissions, energy consumption, water consumption, eutrophication and landfill. This can be captured in a life cycle assessment (LCA), but this method is resource intensive and requires significant expertise to interpret the results. In an LCA, each component needs to be assessed separately. In this chapter, CO_2 emissions are used as the model, as they represent a common measurement for capturing the indicative environmental impact, and because data are available from the underlying research. The CO_2 emissions here can easily be replaced by the energy consumption in the evaluations. The main benefit of focusing on energy is that all energy sources are analysed, both those that generate CO_2 and those that are CO_2 free, such as nuclear and water power. It is therefore possible to identify the total energy efficiencies to be gained. The main benefit of focusing on CO_2 is that the risk of global warming is emphasized. This focus on CO_2 assumes that it is more urgent to reduce fossil-based than non-fossil energy.

Industrial packaging affects many processes, such as purchasing, logistics, production, transportation, warehousing and recycling. With well-informed selection principles for industrial packaging, where multiple trade-offs have been considered and the effects of packaging decisions are known, a more competitive and efficient supply chain can be obtained. It has become increasingly challenging to make well-informed choices of industrial packaging because the complexity of supply chains has increased, because supply chains are increasingly global in nature, and because of reduced lead-times, etc. In addition, in supply chain practices, it is common to focus on the requirements of the channel leader. For instance, in the automotive industry, production requirements have played a decisive role in industrial packaging selection. By incorporating packaging requirements of other functions and channel partners with the existing requirements into a structured analytical approach, however, research and experience show the potentials for both cost and CO_2 reduction because of better-informed trade-offs. This chapter presents an evaluation model which uses such an approach. First, the chapter discusses a generic supply chain and its system boundaries for industrial packaging. With this as a background, it presents and discusses the content of an evaluation model in terms of cost and CO_2 factors. In the final section, it proposes a stepwise development and implementation structure for an evaluation model.

System boundaries for industrial packaging applications

To assess the total costs and CO_2 emissions of packaging and its utilization in a supply chain throughout the packaging life cycle, a system perspective should be applied. It is essential to clarify the system boundaries, which must capture the life cycle of the packaging material as well as the economic and environmental impacts in packaging usage, which ranges from point of filling to point of consumption. Figure 8.1 illustrates the system boundaries of industrial packaging in a generic supply chain. These system boundaries capture the net effect of the costs and the environmental impacts for the packaging material and its recyclability or material recovery. The boundaries also capture the impact of packaging on transportation, storage, material handling, and waste handling or reuse.

Figure 8.1 illustrates a system using both one-way and reusable packaging. Both types of packaging are produced by a packaging supplier, and delivered to a component supplier. The one-way packaging can often be flat, which enables efficient transportation, whilst the reusable packaging requires

Figure 8.1 System boundaries of industrial packaging in a generic supply chain

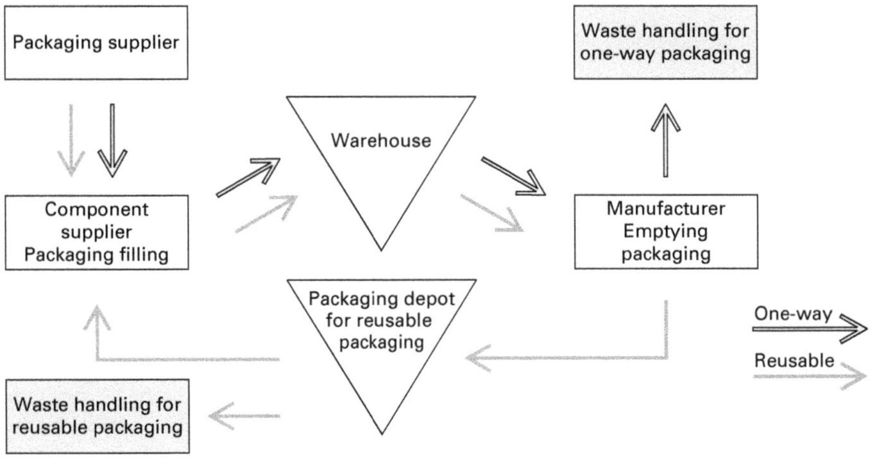

more space, depending on the type of reusable packaging (see Chapter 7). At the component supplier, packaging filling takes place, before the packaging system is loaded onto vehicles for transportation to a warehouse. The filling process involves the entire packaging system. An example would be secondary packaging with components on a pallet with or without collars or shrink-wrapped plastic film. In the warehouse, components or packages may be repacked to form a new packaging system, which is then loaded onto a new vehicle for transport to a manufacturer where the components are used. The empty packages are transported either to a recycling facility (one-way packaging) or to a packaging depot for reuse. After a certain number of usages, the reusable packages are also transported to recycling. These two transports can be different in terms of distance, vehicle type and load factor. The one-way packaging is usually compressed, whilst the reusable is stacked. In the packaging depot, the reusable packaging is washed and then reused.

Cost and CO_2 factors for evaluation of industrial packaging systems

Five factors (packaging material, transport, material handling, waste handling and capital) drive the total costs of industrial packaging in supply chains. In a similar way, four factors (the same factors, minus capital) drive the total CO_2 impact. The performance of the factors is affected

Figure 8.2 Descriptions of factors needed to calculate the impact of packaging on supply chain costs and CO_2

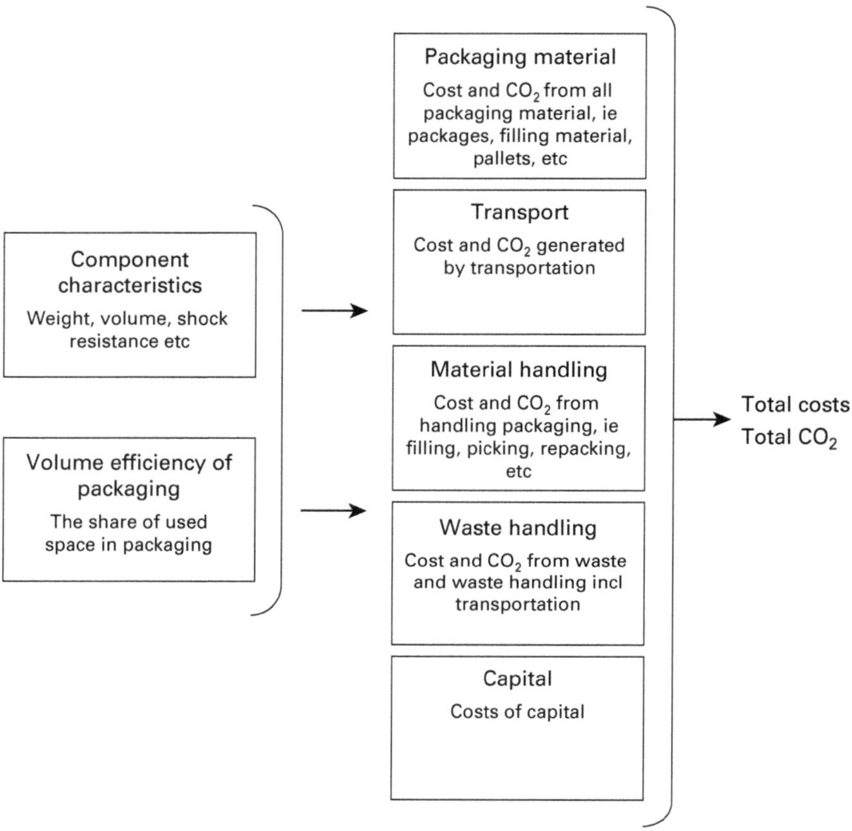

by component characteristics and the volume efficiency of the packaging system. The factors are described in Figure 8.2.

The volume efficiency of the packaging system is considered in each process (eg for material handling and transport efficiency). It determines the share of used space in the packaging system, ie the combined used volume in the primary, secondary and tertiary packaging. If the used space in the primary packaging is 80 per cent, in the secondary packaging 50 per cent and in the tertiary 70 per cent, the total used volume in the packaging system is 28 per cent (0.8 × 0.5 × 0.7). This affects the costs and CO_2 of the five factors in Figure 8.2. The volume efficiency of packaging is similar for both costs and CO_2 environmental and economic criteria, but its value varies for different parts of the supply chain; packaging systems, for instance, may be repacked. Thus, separate measures for the volume efficiency of packaging should be calculated for each part of the supply chain.

The *packaging material* factor covers the costs and CO_2 from all packaging material. The production of packaging requires raw material and resources for working up the material. The packaging material factor includes not only the packages, but the added material needed, such as stretch film or intermediate layers of corrugated board. Other examples of costs and CO_2 related to the packaging material factor are purchasing, handling, storage, as well as repair and maintenance of packaging. For reusable packaging, administration costs related to interest charges for service, distribution and collection of packaging should be included, as well as salaries for managing reusable packaging systems and costs for investments in information systems. Administration costs for one-way packaging are considered in the waste-handling factor, as it refers to collection fees. The packaging material costs of one-way packaging are straightforward, as this goes to waste after disposal. In calculating the material costs for reusable packaging, the number of times the packaging is used before recycling needs to be considered. The calculation of CO_2 from packaging manufacturing should take the share of recycled material into account.

The *transport factor* covers the impact of transportation in several parts of the system under study. In Figure 8.1, transportation takes place on six different routes, such as from the packaging supplier to the component supplier, and from the component supplier to the warehouse. In the figure, the arrows represent a transport, and each transport stage can involve intermodal transport. For example, in global supplies from Asia to Europe, truck transportation to a port can be followed by sea transport, and then another truck transport to a warehouse. In these intermodal transport cases, the cost and CO_2 of each transport should be compiled after individual calculations. To determine the costs and CO_2 of transportation, the following variables are needed: the costs and CO_2 for the transport mode used, shipping distance, empty running, load factor, and equivalent maximum capacity in a vehicle. The costs and CO_2 of empty running should be included if the return transport of the vehicle is not utilized on the return route. The maximum loading capacity of a vehicle helps to allocate the costs and CO_2 of a vehicle to one package (and in the next step to one component). The costs and CO_2 of a vehicle per package is estimated by defining the hypothetical maximum number of packages in a vehicle (if the vehicle only uses that type of packaging system) and then dividing the emissions/costs of a vehicle by this figure. Transportation related to waste handling is evaluated in the waste-handling factor, which means that the costs and CO_2 of return transport of reusable packaging, such as pallets, are calculated there.

The factor for *material handling* captures these activities in manufacturing plants and warehouses. To estimate the costs and CO_2 of material handling, the material flows in plants and warehouses need to be mapped. These flows include activities, such as picking, erecting packaging, movements, repacking, unloading and more. If possible, it is recommended that type flows are mapped and defined, and component (product) and packaging types linked to different flows. For example, the components in packaging of one type are first stored, then moved by a forklift to another area, before they are put on a conveyor belt for transfer to the point of consumption. Components in another type of packaging may be repacked into different packaging, and then put on a tugger train to reach the point of consumption.

As these examples indicate, the costs and CO_2 of each type flow varies. The costs of material handling originate from labour and equipment costs for both manual and automatic handling. The CO_2 of material handling comes from energy needed for material-handling equipment, such as forklifts, conveyor belts or industrial trucks. It also comes from the warehouse space needed for a package. One approach to allocating CO_2 from a warehouse building to packaging is to estimate its share of the total warehouse space. As the aim is to calculate the cost and CO_2 efficiencies of using a particular packaging system for a component, the volume efficiency factor is essential in determining material-handling efficiency. For instance, in manual handling, the time each component is handled is considerably lower for a package with many components than for a package with few components. Another essential consideration for the efficiency of activities in material handling is whether several packages can be handled together. For instance, if a forklift can handle two packaging systems instead of one, its effect per component is halved.

The *waste-handling factor* includes cost and CO_2 for waste, recycling and reuse of packaging systems as well as damaged components because of insufficient packaging. The variables for this factor are different for one-way packaging and reusable packaging. The costs and CO_2 for one-way packaging refers to storage, transportation and recycling of waste. For reusable packaging, the costs and CO_2 refer to storage, and transportation of empty packaging and packaging waste, as well as to cleaning, recycling or energy recovery. The lifetime of reusable packaging should be considered in the evaluations. For example, if a package is used 25 times before it is removed, and recycled the total costs and CO_2 of the waste-handling factor should be divided by 25 for allocation to a component. For reusable packaging, particularly for high-value packaging, there is a risk for loss of packaging because of pilfering: for example, roll cages are quite expensive and have many application areas.

Furthermore, if the recycling or material recovery generates an economic income or CO_2 from positive energy output, it should be considered in the waste-handling factor. Thus, the energy extracted from packaging waste reduces the total energy consumed in the lifetime of the package.

The *capital factor* refers to the capital costs of components in inventory. This cost depends on the components' storage time. In general, the more items that are ordered at any one time, the longer the storage time. If in principle packages must be full before they can be shipped, then larger packaging sizes have a higher minimum order quantity than smaller sizes. Therefore, the average time in storage is longer if the number of components needed to cover an order exceeds the economic order quantity (EOQ), because of the packaging size. For example, assume the EOQ is 10, but the package holds 8 components. If the packages held 5 components, then the exact EOQ could be ordered, ie 2 packages of 5 components. However, with 8 components per package it will take longer to use all 16 components, which increases the average storage time. (Note that the components in this example are considered too heavy to be in one package.) Accordingly, the capital factor can increase with larger packaging, but on the other hand, the transport and the material-handling factors often decrease with larger packaging.

The other factors are also affected by the packaging size. Thus, to evaluate the total effects, it is necessary to calculate the combination of the factors. To do so, the cost and CO_2 drivers of each factor should be considered. These are summarized in Table 8.1. For instance, to calculate the transport cost, it is necessary to know the cost per kilometre for the mode of transport used, the shipping distance, and the kilometres of empty running related to the delivery. In order to describe the transport cost per component (ie the transport cost per component of using a particular packaging system), it is also necessary to know the load factor of the vehicle and the share of the space that the package uses. The latter can be calculated by determining the maximum number of packages in a vehicle of the packaging type that is being analysed. This can be restricted by either the maximum weight or the maximum volume of the vehicle. The share of the costs and CO_2 from the vehicle to a particular package is the share of the package compared to the maximum number of packages in a vehicle multiplied by the load factor. For example, if a truck can fit 48 pallets, and each pallet has 12 packages, the maximum number of packages in the truck is 576. If the load factor is 70 per cent, the share of the truck cost/CO_2 for a package is $1/(70\% \times 576) = 0.25\%$. For comparative reasons, as described above, this value should be divided among components, ie if the package holds 5 components, the transport costs and CO_2 per component are 0.05 per cent (0.25/5).

Table 8.1 Factors and variables needed to evaluate supply chain impacts of industrial packaging

Factor/variable	Cost driver	CO_2 driver
Degree of filling in primary packaging	%	%
Degree of filling in secondary packaging	%	%
Degree of filling in tertiary packaging	%	%
Packaging material		
Packaging material	Cost/kg	CO_2/kg
Packaging material weight	kg	kg
Primary energy use in packaging production	N/A	CO_2/package
Packaging investment	Cost	N/A
Administration (service, distribution and collection)	No. of packages*	N/A
Transport		
Mode of transport	Cost/km	CO_2/km
Shipping distance	km	km
Empty running	km	km
Load factor	%	%
Equivalent max capacity in transport (no. of packages)	No.	No.
Material handling		
Movements in material handling	Cost/h	CO_2/h
Warehouse space needed	Cost/m^2 per year	CO_2/m^2 per year
Sorting	Cost/h	N/A
Packing process	Cost/h	CO_2/h
Waste handling		
Cleaning/reuse	Cost/h	CO_2/h
Loss of packaging	Cost/package	CO_2/package
Recycling	Cost/h	CO_2/h
Shipments of packaging waste	Cost/tonne-km	CO_2/tonne-km
Packaging waste	Cost/kg	CO_2/kg
Product shrinkage	Cost/product	CO_2/product
Capital		
Capital cost of components in inventory	Interest rate, storage time	N/A

*A more comprehensive cost driver combines packaging fleet size, no. of actors, and no. of packaging types

This section has described an analysis structure, and factors to estimate the total impacts of packaging systems on costs and CO_2. The next section elaborates on how these factors can be combined into an evaluation model.

Guidelines for developing a packaging evaluation model

Every time industrial packaging for material supply is used, a unique combination of type of component (product), supplier distance, material handling, storage time, etc affects logistics performance. It is almost impossible to select the most efficient packaging system solution without detailed calculations. In a research study in the automotive industry, an analysis model was developed and applied, based on the factors discussed in this chapter. The model takes packaging material and waste-handling activities into account as well as all activities that are affected by the packaging system from point of filling at a material supplier until point of consumption in a manufacturer's plant. The logic and basic structure of the model were developed in Pålsson *et al* (2013), and then applied and refined by Wallström and Pålsson (described in Carlsson *et al*, 2015). The model has been implemented in a number of plants in the automotive industry, and enables the companies to make informed packaging decisions, both operationally and strategically.

In addition to providing accurate decision-making support for operational packaging selection, an analysis model that calculates the impacts of packaging on costs and CO_2 in supply chains can help in other decision-making. For example, it can help in the following ways: creating guidelines for packaging selection; identifying and managing trade-offs between cost and environmental impacts related to packaging; whether to use repacking or not in the material supply; selecting the most appropriate packaging size; and providing strategic decision support for supply chain management, sourcing and production. More details on the value of a model can be found in the next chapter, which discusses decision support for industrial packaging, first in general terms and then translated into packaging selection principles with illustrations.

The suggestion here for a model that evaluates the environmental impacts of industrial packaging follows the logic of life cycle assessments (LCAs) in that it considers the life cycle of packaging, and its usage from packaging supplier to the packaging end of use and waste handling. However, the proposed evaluation model differs from LCAs, as it includes both economic and environmental features, and only considers CO_2 for the environmental impact. The proposed model can be used to calculate other environmental measures in a similar way, but that requires additional data capture. If different measures are combined, the decision measures become more complicated to interpret. The proposed model should include both the net effect of the costs and CO_2 for the packaging material and its recyclability or material recovery along with the impact of packaging on the ordinary processes of a supply chain at the component supplier, transportation, storage, material handling, and waste handling or reuse (Figure 8.1). The volume efficiency of the packaging system should be considered for each factor, such as for material handling.

The basis of an evaluation model should be to calculate the total cost and CO_2 impacts of using a particular packaging system. The model should take into consideration both the direct impact of the packaging material and the indirect impact of the utilization of packaging in the supply chain. As described in this chapter, the economic calculations should be based on costs, whilst the environmental calculations can be based on, for example, CO_2.

The ability to compare the costs and CO_2 of using different packaging systems opens up several opportunities. Two examples would be to evaluate whether it is more or less costly to use a one-way packaging system compared to a reusable one, or to evaluate which of several different packaging sizes are the most cost and CO_2 efficient. The component is the lowest common denominator in the flow of goods and between different flows. Thus, to capture comparative figures of total costs and CO_2, the calculations should be done per component, ie an aggregated cost or CO_2 figure for a package including more than one component should be divided by the number of components. The same logic should be applied to the other factors, such as transport, material handling and packaging material. Thus, each factor shows the cost and CO_2 per component of using packaging system X. The total cost or CO_2 per component is evaluated by summarizing the cost or CO_2 factors per component. Accordingly, the total cost or total CO_2 of using packaging system X can be compared to the corresponding figures for using packaging systems Y and Z.

Proposed development steps

Much was learnt from a research project that developed and implemented an evaluation model for industrial packaging systems in the Swedish automotive industry (eg Carlsson *et al*, 2015; Pålsson, Finnsgård and Wänström, 2013; Pålsson, Wallström and Johansson, 2014) where Wallström played a key role in the operationalization of the model. Based on this research and the factors in Figure 8.2, the following section proposes an approach to developing an evaluation model for industrial packaging systems.

The initial step is to map and collect basic empirical data about processes to understand supply chain characteristics. With this as a basis, it is possible to follow the analysis structure and the content of factors as described in this chapter to create a general set-up and structure of an evaluation model. Such a general structure should capture and describe process flows, activities, and general product and packaging characteristics. Data that depend on specific contexts should be added later. This can be, for instance, plant-specific activities or certain handling times in a warehouse. For example, during the implementation phase of the research project, specific data from three plants was input into the model to provide context. Each plant had five similar types of material flows, but the order and time consumption of the activities varied between the plants. The transport data were, however, similar in the flows to the three plants. In other contexts, this will be different.

In the research project, a key to successful implementation was to have a continuous dialogue with representatives from all functional areas. This took place in intra-organizational meetings and workshops. It was particularly essential that the logic and structure of the model and the data within it reflected reality. This ensured that the model could be trusted and was applicable to different parts of the organization. An example of this was a collaborative approach to create general descriptions of material flow types. In short, the development and implementation of an evaluation model for industrial packaging can include the following steps:

1. Form a comprehensive intra-organizational project group with representatives from production, purchasing, sourcing, supply chain management, logistics, packaging, environment, IT and transportation.
2. Determine system boundaries including the number of plants.
3. Conduct process mapping and initial data collection to understand supply chain characteristics and material flows.
4. To start with, select one plant and all flows of component supply to this plant.
5. Build a draft version of an evaluation model based on the analysis structure and factors in Figure 8.2 in combination with the initial data collection.
6. Discuss the structure, content, level of details, material flows etc in the draft version of the evaluation model with representatives from the intra-organizational project group.

7 Expand the draft version into a first complete version with ongoing intra-organizational discussions

8 Discuss the first version of the evaluation model in the project group.

9 Update the evaluation model according to the output from Step 6 and add more detailed data if needed. This will result in a second (third, fourth, etc) version.

10 Discuss the second (third, fourth, etc) version of the evaluation model in the project group. If necessary, go back to Step 7.

11 Launch a final, complete version of the evaluation model for the first (second, third, etc) plant.

12 Adjust the analysis model for other plants by following Steps 1–11. This will be a much smoother process than the first iteration.

The key to this procedure is to carry out iterative steps that can be discussed with the project group. This helps to capture sufficient levels of detail in the data and to anchor the evaluation model in the organization, which is necessary in order to utilize its capabilities. These relate to decision support for supply chain practices, which is discussed in the next chapter. It should also be noted that a considerable amount of work is required to structure and collect data about costs and CO_2 from transportation, materials handling, waste management, packaging material, warehouse management and administration. Various data sources and observations are required.

Summary

This chapter focused on industrial packaging and how to evaluate it from economic and environmental perspectives. Industrial packaging's main features are to protect the products, facilitate efficient production, material handling, warehousing and transportation processes. A more competitive and efficient supply chain can be achieved with well-informed selection principles for industrial packaging, where multiple trade-offs are considered and the effects of packaging decisions are known.

The chapter presented an evaluation model that uses a structured approach to evaluate the supply chain performance of industrial packaging from a systems perspective. The evaluation model considers both economic and environmental effects. Initially, the chapter discussed a generic supply chain and its system boundaries for industrial packaging. This is the first step in an evaluation. Then it presented and discussed the content of an evaluation model in terms of cost and CO_2 factors.

The evaluation model includes five factors (all of which affect cost and four CO_2) that drive the total costs and CO_2 of industrial packaging in supply chains: packaging material, transport, material handling, waste handling, and (a cost factor) capital. The performance of these factors is affected by component characteristics and the volume efficiency of the packaging system. The *packaging material* factor covers the costs and CO_2 from all packaging material. The *transport factor* covers the impact of transportation in several parts of the system under study. The factor for *material handling* captures costs and CO_2 activities, such as picking, erecting packaging, movements, unloading, and repacking, in plants and in warehouses. The *waste handling* factor includes cost and CO_2 for waste, recycling and reuse of packaging systems as well as damaged components because of insufficient packaging. The *capital* factor refers to the capital costs of components in inventory.

The chapter went on to discuss guidelines for developing a packaging evaluation model to get useful outputs. It is almost impossible to select the most efficient packaging system solution without detailed calculations. This is because for each usage of industrial packaging for material supply, a unique combination of type of component or product, supplier distance, material handling, storage time, etc affects the logistics performance.

Finally, the chapter proposed a stepwise development and implementation structure for an evaluation model. The intention was to propose an approach for developing an evaluation model for industrial packaging systems based on what was learnt in a research study.

Decision support for industrial packaging 09

This chapter discusses decision support and selection principles for industrial packaging. It also emphasizes the possibilities of such principles by illustrating their effects on costs and CO_2 in a number of real-life examples. As described in the previous chapter, the economic and environmental performance of industrial packaging can be determined from five cost factors: and four CO_2 factors packaging material, transport, material handling, waste handling and capital costs (only a cost factor). The combined effects of these five (four) factors, which show total supply chain costs and CO_2, need to be considered in order to provide guidance and to formulate selection principles for industrial packaging.

Decision-support areas from industrial packaging evaluation

As indicated in the previous chapter, the characteristics and set-ups of industrial packaging (eg size, reusable or one-way packaging, full or EOQ on pallets, repacking or not etc) in processes for supply of material affect supply chain cost and CO_2. It is a complex matter, though, to determine the total cost and total CO_2. This is because it is a combination of direct and indirect costs and CO_2 related to the five (four) factors: packaging material, transport, material handling, waste handling, and the cost of capital (described in Figure 8.2). Chapter 8 describes an analysis structure and a model for calculating and analysing the combinations of these factors. By applying such an approach to industrial packaging on an aggregated level, a company can generate decision support for industrial packaging and for the supply of materials. These kinds of decision support are useful in day-to-day business operations to help calculate the costs and environmental impacts of industrial packaging solutions for a new component, or to deliver a requested change to a packaging solution or material flow for a component. They are also useful for establishing selection principles for industrial packaging. This section explains which kinds of decision support and analyses can be developed from these holistic evaluations of industrial packaging, and also describes how the evaluations can provide input to guidelines for designing material-supply processes.

An evaluation model containing the cost and CO_2 factors can generate decision support in several areas. First, it can emphasize the relative importance of each cost and CO_2 factor. This helps to understand factors with high impact (ie where to reduce costs and CO_2), and makes it possible to determine if it is worthwhile to put efforts into making a given factor more efficient. For instance, if an evaluation shows that the material-handling costs for a component represent 40 per cent of the total cost, there are significant savings to be made if material-handling efficiency can be increased. However, if it only represents 5 per cent of the total costs, the savings potential would be quite low. In that case, it would be more valuable to focus improvement efforts on other factors. Similar examples can be presented with CO_2.

Second, an evaluation model can calculate the total costs and total CO_2 in a supply chain from a life cycle perspective. By making such an analysis per component, for groups of components or for all components, it is possible to understand where costs and CO_2 arise in a supply chain. After creating a reference point with the existing industrial packaging systems, comparable calculations for contemplated systems can be made. This will show the cost and CO_2 differences between various packaging systems. One specific, strategic comparison is between one-way or reusable packaging systems. The individual cost and CO_2 factors of these two industrial packaging systems usually vary. By evaluating the total costs and CO_2 factors, insights into the supply chain efficiency can be gained, and factors needing most attention in each packaging system highlighted.

Third, after having determined whether to use one-way or reusable packaging, or a combination of the two, an evaluation model can assess different packaging alternatives. It can be applied to estimate the costs and CO_2 impacts of the volume utilization of packaging. For instance, if the volume of the packaging system is not utilized to its full potential, or if a slight change in dimension in the packaging system would increase the number of components more than proportionally, the total impact can be established. The model can also be applied to identify the costs and CO_2 impacts of using different packaging sizes. In that case, the value of some factors will increase, while others will decrease. The model can show the combined effect, which usually is not intuitive.

Another area of decision support is whether to use pallets or containers. By changing the number of secondary packages on pallets or in containers, insights can be gained into the total supply chain efficiency of having full pallets or less-than-full pallets. Full pallets increase transport and material-handling efficiency, whereas less-than-full pallets reduce capital costs, especially for low-volume components, as these have a lower turnover. In some situations, companies want to maximize transport efficiency by using bulk transport, for example, by utilizing pallets with collars without secondary packaging. However, smaller packages may still be desired in the manufacturing plant, and this requires repacking. An evaluation model helps to decide whether to use bulk transportation with repacking into smaller packaging, or to deliver small packages from the point of filling.

Additionally, an evaluation model can offer decision support for scale effects. At a cut-off point, individually calculated packaging solutions may not be the most cost or CO_2 efficient when they are combined. For instance, a component supplier may need to store and handle several packaging sizes, but it may be more cost efficient to store only a few. Thus, an evaluation of the total scale effects for a supplier, a region, or the number of packaging options supports decision-making. Note: the scale effects can be integrated with comparisons between one-way and reusable packaging in order to identify different packaging alternatives in both.

An evaluation model based on these factors can also be used to challenge and understand the impacts of well-established and overriding principles governing material-supply decisions. This is because these principles influence industrial packaging selection, which in turn influences costs and CO_2. The guiding principle can be that packaging size should cover a few hours of production time, optimize the space utilization in the production process, minimize total supply chain costs, or minimize total CO_2 in the supply chain. Each of these principles usually results in four different packaging solutions with four different cost and CO_2 impacts. A guiding principle can result in different packaging sizes and repacking, and this would affect a majority of the factors.

Evaluations taking into account the five cost and four CO_2 factors contribute primarily to decision support for industrial packaging from a holistic perspective. However, they can also increase cross-functional and inter-organizational collaboration to reduce supply chain cost and CO_2. The indirect, positive effect of working with industrial packaging as proposed is that this facilitates cross-functional and external collaboration. It does so because of the need to make trade-offs between requirements from multiple functions and supply chain actors. The many requirements range from component suppliers, via transport providers to warehouses, plants, and stakeholders in recycling. Collaboration is necessary in order to make informed trade-offs. This effect should not be overlooked.

Selection principles for industrial packaging

Packaging selection principles are the basic rules for selecting packaging. These principles should provide guidance on how to minimize and manage trade-offs between supply chain costs and the environmental impact of packaging. They should also consider ergonomic aspects and practical constraints (eg space limitations), but the focus here is on the economic and environmental impacts, as explained in Chapter 8. The ergonomic aspects and practical constraints can be included in the selection principles as end conditions. From economic and environmental perspectives, the packaging selection principles should go beyond the direct and indirect effects of packaging to include the effects of how the material supply is planned and organized as well. There is considerable cost and CO_2 reduction potential in the direct and indirect effects of packaging, along with the effects from material supply, which will be illustrated in this chapter. This section explains a number of selection principles for industrial packaging. The sections that follow present real-life examples of the principles.

Sourcing and production strategy decision principles

One decision related to industrial packaging is primarily based on sourcing and production strategies, and how they are operationalized. This has considerable impact on the efficiency of material supply because the strategies guide order frequency, order size and lead-time. As a result, two options for guiding the material supply are: minimizing total costs or minimizing total environmental impact. A third option can be that an order should cover a certain period of production time. A fourth, less holistic option, is to minimize inventory costs. As a fifth option, a part of the material supply guidance can be to minimize packaging material costs. There are also other possible ways to guide the supply of material. However, the message is that the aim for efficiency in material supply and production affects packaging efficiency. For instance, if the aim of material supply is to support just-in-time deliveries with minimal inventory costs, certain packaging solution characteristics are encouraged, whereas if the aim is to minimize the total costs of packaging material, transportation, inventory and material handling, other characteristics of the packaging system solution are preferable. In the first example with just-in-time deliveries, quite small packaging sizes support frequent and small order quantities and do not have to be repacked or result in inventory. In the total cost example, a similar packaging solution could be the most efficient, but if the transport distance is long, it could be more cost efficient to use bulk transportation with repacking into smaller packages downstream in the supply chain. It is a complex matter to determine the most cost-efficient packaging solution because the total cost perspective has five factors to take into account. Thus, the proposed model can be useful.

One-way or reusable packaging decision principles

A second decision for industrial packaging is whether to use one-way or reusable packaging, or a combination. The combination of a number of contextual factors, such as transport distance, the types of material flows in plants, and the product characteristics, all determine the total supply chain effects of packaging. The existence of an industry consortium for reusable packaging and an influential supply chain leader also promotes reusable packaging, whereas the presence of one-way packaging suppliers facilitates such utilization. The product characteristics can affect the weight and volume efficiency of one-way or reusable packaging differently. Heavy products may reach the maximum handling weight before a reusable package is full, because it is heavier than the one-way alternative. On the other hand, the one-way package may not be strong enough to handle heavy products. In a similar manner, the shape of certain products can utilize additional inner volume in a one-way package with rectangular shapes compared to a conical reusable package, whereas other products cannot. Thus, in order to select when and where to use one-way and reusable packaging, the cost and environmental impacts of contextual factors and product characteristics need to be analysed for the particular choice.

Packaging size decision principles

A third decision for industrial packaging that can be both strategic and operational is the packaging size. An initial consideration is the daily and annual consumption of the packed component. To a certain degree, the lower the consumption, the smaller the packaging size. For instance, a big package can result in unnecessary inventory and space utilization in plants. On the other hand, when the annual consumption exceeds a number of full pallets for various packaging sizes, other supply chain characteristics must be considered.

Packaging size affects efficiency in picking, transportation, handling, etc. It also affects the effectiveness and efficiency of the manufacturing process when the components are picked for assembly or refinement. For instance, a small package may be more accessible and free up space for other components on an assembly line. However, the size is not only about the total volume, but also about the shape. A wide and long package may be more accessible and require less picking time to empty than a high package with short sides. An alternative approach in the selection of packaging size is to postpone the final packaging solution by using repacking. Figure 9.1 presents an example where the number of components per pallet is 20 per cent higher in the pallet with collars (1,500 components) compared to the ready-packed components (1,200). However, components in the pallet with collars need to be repacked before entering the production plant. If the transport is restricted by volume, the pallet with the collar alternative is approximately 20 per cent more cost and CO_2 efficient (the fuel consumption per truck increases slightly because of increased weight, but not in proportion to the additional components), but there is a trade-off for an additional repacking cost. Thus, if transport distance A in Figure 9.1 is long enough so that improved transport efficiency outweighs the additional repacking, the pallet with the collar alternative is the most cost efficient. From an environmental perspective, this alternative is likely to be most efficient irrespective of repacking efficiency, particularly if the repacking is done manually.

Scale effects decision principles

A fourth decision is based on the potential scale effects. It is possible to identify economies (and the environmental impacts) of scale for packaging solutions. Analyses of packaging solutions for individual components, for instance, can result in a vast number of packaging solutions for a certain supplier or transport provider, or in a geographical area. This means that the individual analyses show that one package type is the most beneficial for one component, another package type is the most beneficial for another component, a third package type is the most beneficial for a third component, and so on. However, the scale effects may show that it would be beneficial to use fewer unique packaging types for a supplier, a geographical area, etc. Having many packaging types can cause inefficiency in material handling or transportation because of co-loading challenges (unless different packaging solutions are modularized). It may also require unnecessary safety stock, because a supplier and a geographical area need to hold inventory for each packaging type it uses in order to address uneven demand. Having additional

Figure 9.1 Illustration of cubic utilization with and without repacking

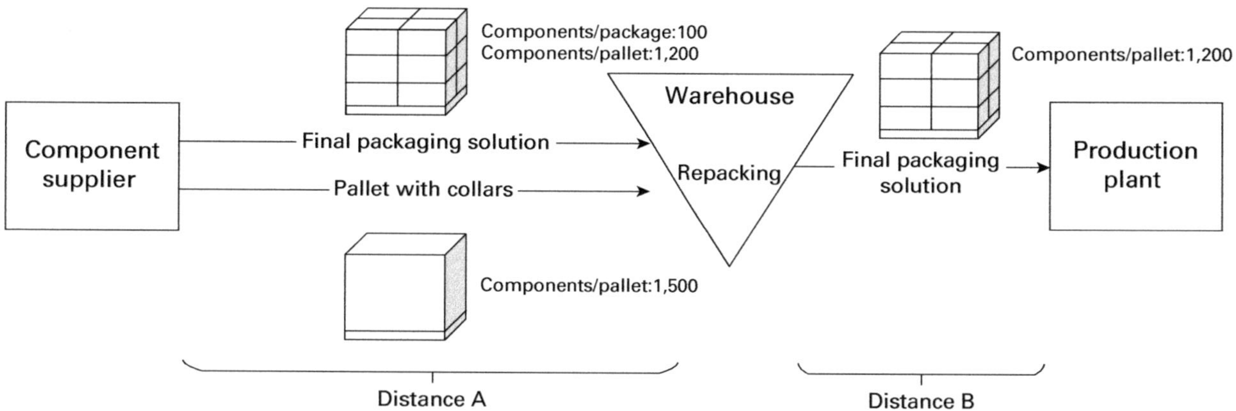

safety stock for each packaging type increases the total fleet size. More types of packages also increase the complexity and cost of administration.

To consider the scale effects, the total supply chain costs and CO_2 effects of modifying the individual packaging solutions needs to be analysed. One proven analysis approach involves the following steps:

A Identify a number of suppliers or a geographic area to analyse.

B Determine the number of different packaging types used at those suppliers or in the geographic area.

C Identify packaging types used for high-volume components.

D Reduce the number of packaging types.

E Analyse total cost and CO_2 effects.

F Compare cost and CO_2 effects to potential savings in material handling, administration, fleet size and safety stock.

The analysis should start by identifying the number of suppliers or a geographic area to analyse. Then, the number of different packaging types used at those suppliers or in the geographic area should be determined. This is followed by identifying the packaging types with the biggest volumes. Thereafter, the least-used packaging types can be changed to the nearest size that is kept in the possible scenario of a new configuration of packaging types. To decide whether a new packaging configuration should be used, the total supply chain costs and CO_2 of the effects of the changed packaging types should be calculated. These effects should be compared to potential savings in material handling, administration, fleet size and safety stock. In some cases, the scale effects show savings, but in other cases, the original packaging types are the ones that are most cost and CO_2 efficient. In addition, there is not a definite value for the number of least-used packaging types in Step D. Therefore, a number of different analyses using Steps D–F should be carried out. In these analyses, the unit of analysis can first be suppliers and then a geographic area.

Volume and weight utilization decision principles

Finally, the fifth decision concerns the volume and weight utilization of the packaging system. Here, it refers to utilizing the capacity of the packaging system. For instance, the order quantity may show that less-than-full secondary packages or less-than-full pallets should be supplied. For components where the order quantity results in less-than-full packages, the economic and environmental effects in the supply chain of filling up the secondary packaging and the tertiary packaging (eg pallets) should be analysed. By filling up packages, the costs for material handling, transportation and to some extent the packaging material are reduced, whereas the risk of obsolescence and capital cost of inventory increases. The total effects need to be calculated for each context.

Illustrations of industrial packaging principles

As described above, different principles for selecting industrial packaging lead to different cost and CO_2 impacts. The previous section presented five decisions for selection principles for industrial packaging. To support these decisions, detailed analyses are needed. Each decision is illustrated here by applying it to a case of material supply in the automotive industry. The illustrations include:

- sourcing and production strategies;
- one-way vs reusable packaging;
- packaging size;
- scale effects;
- volume and weight utilization.

This illustration of cost and CO_2 impacts of selection principles for industrial packaging is based on material supply in the automotive industry. The figures have been modified to ensure confidentiality, but the information is based on a case study about the Swedish automotive manufacturer Volvo and its European component suppliers. To a great extent, the components are packed in reusable packaging in these material flows. The supply of components comes from various European countries to a warehouse in Gothenburg where they are unloaded and then transported to a production plant. The warehouse is also a packaging depot where empty packages are washed and stored. In the production plant, various material-handling activities occur. Pulling carts move palletized packages to a supermarket area where packages are loaded on flow-racks. Tugger trains are used to move packaging into the production area where empty packages are replaced by full ones. The used packages are sent to the packaging depot for cleaning and reuse, or compressed and sent to a recycling site. The packaging system includes one-way or reusable packaging, such as pallets, lids, middle layers and sleeves.

Sourcing and production strategies

In the production processes, the weekly consumption rate of a component and space restrictions can determine the packaging size. This means that overly frequent packaging exchanges should be avoided in production, which indicates the use of a fairly big package. Space restrictions, on the other hand, indicate a fairly small package. The combination of the consumption rate and the space restrictions results in a requirement for a certain number of packaging components used in production. To fulfil this requirement, components can be supplied in a large package that is then repacked into smaller ones for use in production. Another option is to supply components in smaller packages to be used in production without repacking. In the repacking option, the costs and CO_2 are lower in transport and material handling, but higher in the repacking process. The costs and CO_2 of the packaging material also vary. To estimate the net effect of these differences, detailed calculations are often needed.

Two examples from the automotive industry illustrate the need to incorporate the context characteristics in predicting the total impact of repacking. The examples also highlight the need to decide carefully whether to use repacking or to use the size of packaging already in the filling process.

The first example involves a component that can be supplied either in a packaging system with a pallet (1,225 × 820 mm) and four secondary packages (600 × 400 × 200 mm) per layer or a pallet and eight secondary packages (400 × 300 × 200 mm) per layer. The first option is repacked into the smaller packaging when it reaches the plant, while the second option is stored and then used in the production. Table 9.1 shows the costs per factor. In total, the costs

Table 9.1 Costing structure for two alternative packaging systems with and without repacking for component Y (€/component)

Factor	Repacking*	No repacking
Original package (mm)	600 × 400 × 200	400 × 300 × 200
Package in production (mm)	400 × 300 × 200	400 × 300 × 200
Packaging material (€)	0.06	0.08
Transportation (€)	0.27	0.32
Material handling (€)	0.21	0.19
Waste handling (€)	0.05	0.03
Capital (€)	0.04	0.03
Total (€)	*0.63*	*0.65*

* Repacking from a transport package to a smaller one

are in the same range. The repacking alternative costs €0.63/component, while the alternative, where the smaller package is used from the point of filling, costs €0.65/component, which is a cost increase of 3 per cent per component.

The different cost is explained by the combination of the factors. The material-handling cost is reduced in option 1 due to the elimination of repacking. In many cases, added labour costs from repacking can be decisive for the total cost difference; but in this example the repacking activity was time efficient and the material-handling factor was also reduced in the storage area. The reduction in material handling was because the larger packages required more internal handling in the storing process. Another difference was that there is more packaging material per component when smaller packages are used. Furthermore, the transportation of components in larger packaging is more volume efficient. However, an opposite effect is that in the repacking option, both types of packaging need a return transport. The net effect of transportation was in favour of the repacking alternative. The use of two types of packaging also increased the waste handling in the repacking alternative. Finally, the capital cost increased slightly because of longer storage time, since twice as many components were ordered. Even though the cost of capital doubled, it only represents a negligible share of the total cost for this component.

A second example of repacking or not concerns another component. In production, this component with low weekly consumption was originally packed in a package with the dimensions 600 × 400 × 200 mm. However, in the filling process, a half pallet with collars was used. Accordingly, repacking was required in the plant. After analysing the production space, it was found that the size of a pallet would fit in the production process. To utilize the transportation better, it was decided to test a full pallet with collar without repacking. This resulted in the total costs being less than half, with a cost reduction of 53 per cent per component (Table 9.2). The cost difference

Table 9.2 Costing structure for bulk transport with pallets (with and without repacking) for component X (€/component)

Factor	Half pallet and repacking €	Full pallet in production €
Packaging material	0.15	0.14
Transportation	0.06	0.05
Material handling	0.50	0.13
Waste handling	0.02	0.01
Capital	0.01	0.02
Total	*0.74*	*0.35*

is mainly explained by the reduced material-handling costs from eliminating repacking, which in this example was the main cost driver. Capital cost increased slightly because of longer storage time, since twice as many components were ordered. Even though the cost of capital is doubled, it only represents a negligible share of the total cost for this component.

These two examples illustrate the need to analyse repacking and other packaging changes in material-supply processes, because the cost impact from similar changes can have totally different results. In the first example, repacking was 3 per cent cheaper while in the second example it was approximately twice as costly.

One-way vs reusable packaging

A major decision in component supply is whether to use one-way or reusable packaging systems. This illustrative example is based on data from Pålsson, Finnsgård and Wänström (2013) where both alternatives were tested in the same material flow. The material supply of a component from a European supplier to a plant in Sweden begins with stacked pallets in a truck transport, which runs about 3,000 km. In the plant, the packaging systems are unloaded and moved to a supermarket area, and then stored, before being moved into production. Used packaging is sent either to a packaging depot for cleaning and reuse (reusable packaging) or to recycling (one-way packaging). Using a reusable package in plastic with the outer dimensions of 607 × 407 × 215 mm resulted in a cost of €0.67/component (Table 9.3). If the same component was supplied in a one-way package of similar dimensions, the total cost was €0.39/component. A main difference between these two packaging systems was the inner volume. Because the reusable packaging has conical sides, the inner volume of the one-way alternative was

Table 9.3 Comparisons of cost factors for supplying a particular component in one-way or reusable packaging 3,000 km (€/component)

Factor	One-way packaging €	Reusable packaging €
Packaging material	0.06	0.02
Transportation	0.13	0.24
Material handling	0.15	0.33
Waste handling	0.02	0.06
Capital	0.03	0.02
Total	*0.39*	*0.67*

considerably higher: in fact, in this case almost 20 per cent higher. This had a positive impact on both transport and material-handling efficiency. On the other hand, it led to slightly higher capital costs. Additionally, the packaging material cost per usage was higher for the one-way packaging. However, in this particular case, the improvements in transport and material-handling factors outweighed the increased costs in capital and packaging material. It should be noted that this held true for this particular case only. In other cases, the reusable alternative can be more cost efficient, and it must be stressed that the context of each particular case must be analysed individually.

In a similar fashion, Table 9.4 shows cost and CO_2 per factor as well as totals. In this case, the total CO_2 was also more efficient in one-way packaging because of transport efficiency (ie increased volume utilization of the packaging system). The other CO_2 factors increased, but not enough to recommend the reusable packaging system.

As noted, each particular supply chain context, in combination with the characteristics of the packaging system and the component, needs to be analysed to identify whether a one-way or a reusable packaging system is the most cost and CO_2 efficient. This is illustrated by Figure 9.2, which shows the cost difference between using a reusable and a one-way packaging system for a number of components from European suppliers to the plant described above. In this example, the current set-up of material supply and packaging size is used for both one-way and reusable packaging. The dimensions are the same for both packaging types. On average, one-way packaging was 7 per cent more expensive. As Figure 9.2 shows, though, reusable packaging is the most efficient option for many components, a considerable proportion of the components are almost equally cost

Table 9.4 Comparisons of CO_2 factors for supplying a component in one-way or reusable packaging 3,000 km (CO_2 gm/component)

Factor	One-way packaging CO_2 gm	Reusable packaging CO_2 gm
Packaging material	32	14
Transportation	314	383
Material handling	—	—
Waste handling	0	16
Total	*346*	*564*

NOTE Electricity from water power (CO_2 zero) in material handling, the sum of waste handling incl energy recovery for one-way packaging was low

Figure 9.2 Cost difference between reusable and one-way packaging system for a number of components

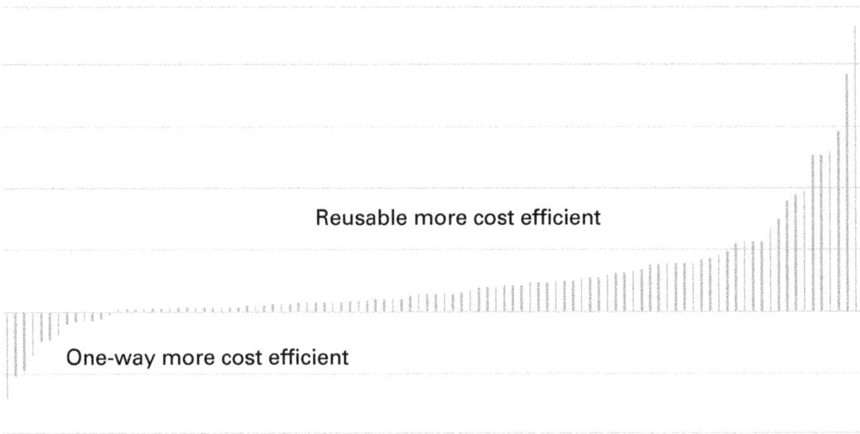

efficient in both reusable and one-way packaging, and some are the most cost efficient in one-way packaging.

Size of the packaging system

As discussed in the repacking example, the size of packaging affects supply chain efficiency. To illustrate the potential impact of only changing packaging size without having to add repacking, consider an example with packaging twice the size. The current packaging system contains 12 packages with 28 components per package on a pallet. The pallet with packages is used in production. If packages of twice the size are used on a pallet, there are 6 packages per pallet. Each double-sized package contains 60 components per package, which means that the number of components increases by more than twice per package (60 vs 28) and by 7 per cent per pallet [(6 × 60) / 12 × 28)]. In this example, the costs are approximately 20 per cent lower through using the larger packaging alternative. The capital costs increase, but are outweighed by the other cost factors. The material-handling factor in particular decreases because fewer packages need to be handled in production. The total CO_2 also decreases by 5 per cent, mainly because of less packaging material per component and increased transport efficiency.

Scale effects

The effects of scale on material supply costs and CO_2 can be illustrated by the way in which instructions from Volvo to their top five European components suppliers are coordinated. In terms of packaging size, for example, it is not just the dimensions of the packaging that are specified, but also whether the pallet is full or not. It is common that the order size in combination with a defined packaging solution determines the number of packages on a pallet. Thus, a pallet can be shipped with less than a maximum number of packages. If a pallet (1,225 × 820 mm) has two packages of 400 × 600 mm, there is space for two more on that layer. There can also be space for more packages on the top. Accordingly, the volume efficiency of the packaging system can be improved if the weight restrictions of the vehicles allow it. From a cost perspective, the factors are affected in a similar way as described for the packaging size.

In order to avoid the risk of having too much inventory, which can result in obsolescence, the number of packages on a pallet was cautiously increased: each layer on a pallet that was not full but had at least one package was filled up. This means that even if more height space were available on the pallet after this calculation, it was not considered, ie if a layer was full then no packages were added even if it was possible to have an additional layer. The results of this calculation showed that these suppliers could reduce the number of pallets from 6,000 to 4,000 per year. They could also reduce the material-handling time by 1,000 hours at the production site, because of fewer pallets to handle. The total cost savings were bigger than the capital costs increase. From an environmental perspective, it was beneficial.

Summary

This chapter discussed and illustrated the usefulness of decision support and selection principles for industrial packaging. It accentuated the need to consider the combined effects of the cost and CO_2 factors (see Chapter 8), which show total supply chain costs and CO_2, in order to provide guidance and to formulate selection principles for industrial packaging.

Decision support from a packaging evaluation model is useful for establishing selection principles for industrial packaging. However, it is also useful in day-to-day business to calculate the costs and environmental impacts of new industrial packaging solutions for a new component, or arising from requests to change a packaging solution or material flow for a component.

This chapter explained which kinds of decision support and analyses can be developed from the utilization of a packaging evaluation model, as suggested in Chapter 8. First, it can emphasize the relative importance of each cost and CO_2 factor. Second, an evaluation model can calculate the total costs and total CO_2 in a supply chain from a life cycle perspective. Third, after having determined whether to use one-way or reusable packaging, or a combination thereof, an evaluation model can assess different packaging alternatives, such as comparing different packaging sizes, using pallets without secondary packaging and applying bulk transportation with repacking or not. The chapter also showed how an evaluation model with these factors can be used in the opposite direction, which means that it can provide input to guidelines for designing material supply processes. In other words, the guiding principles for material supply processes have an impact on industrial packaging selection; examples of such principles are that packaging size should cover a few hours of production time, optimize the space utilization in the production process, minimize total supply chain costs, or minimize total CO_2 in the supply chain.

Packaging selection principles can provide guidance on how to minimize and manage trade-offs between supply chain costs and the environmental impact of packaging. The chapter discussed five decisions that can be covered by packaging selection principles. One decision that impacts the efficiency of material supply is primarily based on sourcing and production strategies, and how they are operationalized. The impacts depend on strategies that guide the order frequency, the order size and the lead-time. A second decision is whether to use one-way or reusable packaging, or a combination. A third decision that can be both strategic and operational concerns packaging size. A fourth decision is based on the potential scale effects, because of the economies (and the environmental impacts) of scale for packaging solutions. The fifth decision concerns the volume and weight utilization of the packaging system.

The final part of the chapter illustrated five decisions for industrial packaging that can be covered by packaging selection principles. The decisions were applied to a material supply case in the automotive industry. This highlighted the complexity and potential benefits of using an evaluation model.

Organizing packaging development

10

The way in which packaging development is organized determines a company's ability to integrate various requirements into the packaging system solution. This affects the performance of packaging in supply chains. Different organizational structures for packaging development facilitate the integration of packaging requirements in different ways. This chapter presents four generic organizational structures for packaging development. It discusses the characteristics as well as the pros and cons of each structure, highlighting areas for particular attention. The chapter also discusses the benefits of concurrent product and packaging design.

Organization of packaging logistics in companies

Research has demonstrated the value of applying an integrative approach to packaging development. As introduced in Chapter 1, supply chain integration is vital for packaging logistics. There are several reasons for designing and developing products and packaging simultaneously:

- By integrating products and packaging systems, it is possible to obtain supply chain efficiency regarding processes in transportation, logistics, manufacturing, and waste management.
- It is also possible to achieve supply chain effectiveness (ie obtaining consumer value). Consumer value can be obtained by avoiding the sub-optimization of products or packages and by focusing on packaging functions perceived to be valuable in a particular context.
- By integrating products and packaging systems, it is possible to obtain innovative solutions. For instance, packaging can be part of consumer

value (ie the packed product rather than the product that is sold). Examples of innovative packaging features for the packed product are promotional attributes, convenience and ease of consumer handling.

An integrative packaging development approach involves more than packaging and product development. It includes developing primary, secondary and tertiary packaging as a unitized packaging system, integrating packaging development with product development, logistics and other functional areas in a company, and involving customers and suppliers in the packaging development process (Molina-Besch and Pålsson, 2013). If the product development team is flexible about product dimensions, and the packaging development team is open-minded about packaging requirements, there is a sound basis for an integrated approach.

An example of how product characteristics can limit the integrated development of the primary, secondary and tertiary packaging is when marketing requirements take priority over other requirements for primary packaging. This can result in the primary packaging being developed in isolation. An example of the lack of integration of various functions is when packaging is considered a commodity or when its role for supply chain effectiveness and efficiency is not emphasized. Then, the functional demands on the packaging may not be strong enough to motivate a functional area to be involved in its development. A final example considers the integration of external supply chain actors in packaging development. If suppliers or customers think that packaging affects their business performance, and can translate this into requirements for the packaging system, it facilitates their engagement in packaging development.

Four generic structures for organizing packaging development

Packaging development can be centralized or decentralized in an organization. In either case, the packaging development can be organized in functions or as projects. The organization of packaging development can thus be described by four generic structures (Figure 10.1). The basic packaging development characteristics of each type were described by Klevås (2005).

In the first structure (top left quadrant of Figure 10.1), packaging development is based on functional and centralized organizations. Packaging development is a support function that develops packaging systems for the

Figure 10.1 Four generic structures for organizing packaging development

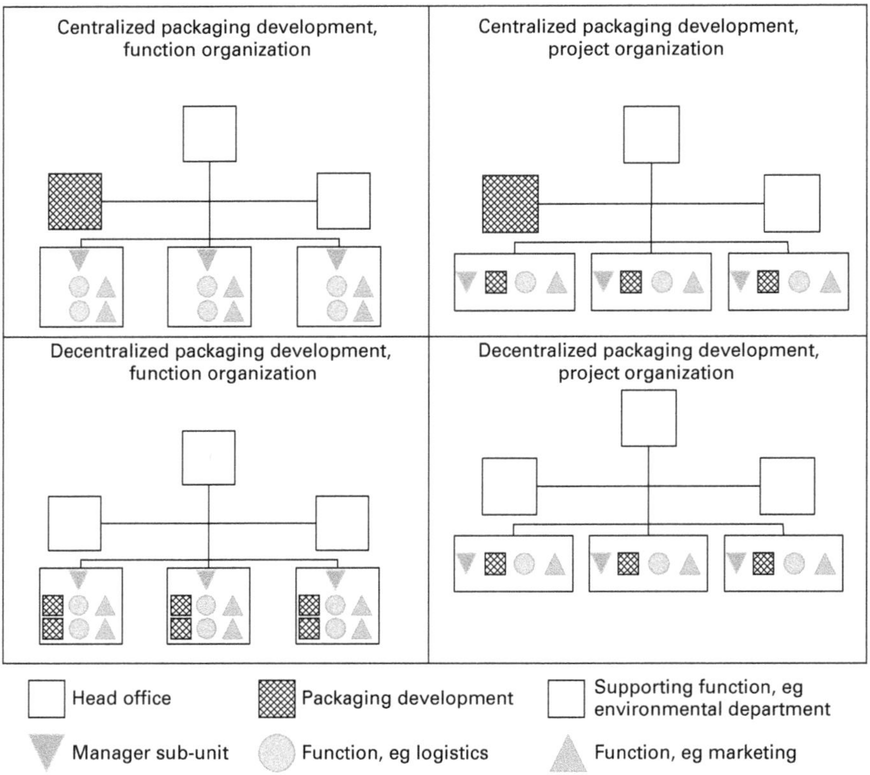

sub-units based on their requirements. The basic packaging development characteristic of this structure is the enablement of high packaging competence and packaging authority. It is linked to logistics, but it is common to have long lead-times for developing packed products (Klevås, 2005).

The second structure is a functions-based, decentralized structure (bottom left quadrant). Here, packaging is developed in sub-units without a centralized coordinator. Packaging development in this structure is closely linked to product development, but lacks clear links to logistics. It is common that the packaging authority is weak in the company as a whole, but strong in the sub-unit. There is an apparent risk of losing packaging competence, as it is distributed among different sub-units.

The third structure is a project-based and centralized one (top right quadrant). This means that packaging development is primarily carried out in a centralized support function, but there are packaging resources in the sub-units as well. Due to the centralization, this structure has high packaging authority and clear links to other functions, such as logistics and product development. Since there are packaging development activities both centrally and in sub-units, the structure facilitates control and coordination of packaging activities, but this comes at a price: there is a risk of bureaucracy.

Finally, in a project-based, decentralized structure (bottom right quadrant), all packaging development activities are conducted in the sub-units. In this structure, packaging development has close links to local customers and product development, but lacks clear links to other areas, such as logistics. The lack of centralized packaging support reduces the risk of bureaucracy, but there are risks of weak packaging authority in the organization and loss of packaging competence. The lack of centralized packaging support also makes it challenging to control and coordinate packaging development between sub-units.

To sum up, these four generic organizational structures have different pros and cons. In practice, organizations use combinations of these generic structures. Thus, it can be beneficial to elaborate on the scope of each of these structures. A thorough discussion about the characteristics and associated pros and cons of centralization vs decentralization and function vs project highlights the embedded trade-offs of each organizational choice.

Next, centralized and decentralized packaging development models are discussed, followed by function- and project-based packaging organizations.

Centralized and decentralized organization for packaging development

To identify areas for particular attention in centralization and decentralization, this section discusses the characteristics, the pros and cons, and the embedded trade-offs of the organizational choice.

The characteristics of packaging development structures based on centralization and decentralization can be grouped into the following five categories.

1 Common packaging principles in a company

This refers to having the same prioritization of packaging requirements throughout a company. It is often easier to focus on the whole company in a centralized packaging development organization, while a decentralized organization more easily identifies itself with the sub-unit. In a centralized packaging development organization, it is easier to align the prioritization of packaging requirements for different business areas or projects based on a holistic overview and better coordination possibilities. Packaging requirements that may need to be aligned include how to prioritize between economic and environmental requirements, or how to measure and align goals for different packaging features. In a decentralized packaging development organization, it is more challenging to coordinate the decentralized packaging decisions. For instance, one decentralized unit can focus on packaging protection, while another on packaging cost and a third on environmental concerns. To align packaging development principles in decentralized organizations, there is a need for a synchronized, overall strategic packaging development process. Otherwise, there is a risk of sub-optimization, as each department or business unit may prioritize internal costs related to the decentralized packaging development over the overall total economic aspects of the company or supply chain.

2 Management and division of packaging development resources

Whether packaging development is centralized or decentralized has an impact on the organization's ability to manage and divide packaging development resources between different product categories, projects and business units. A centralized packaging development organization enables well-planned resource allocation across the entire company, and ensures that management attention for various organizational parts is based on informed decisions. The allocation of resources for packaging development should support the overall performance. In a decentralized packaging development organization, it is more challenging to prioritize between different packaging projects and to have a complete overview of the packaging development, but this enables more rapid response to changed market requirements and customization of packaging.

3 Capturing, maintaining and disseminating up-to-date packaging knowledge

The degree of centralization affects the information capture needed to enhance knowledge. A centralized packaging development organization generally has a greater market scanning capability because more resources can be concentrated on market scanning to identify novel packaging insights. However, a decentralized organization can capture local market information for incremental packaging development due to its local presence. The degree of centralization also affects the capability of maintaining packaging knowledge within the company over time. Packaging knowledge is generally stable over time in centralized organizations; it is not dependent on certain individuals. In decentralized organizations, it can be challenging to maintain knowledge over time, as it may depend on a few key resources. Finally, the degree of centralization affects the dissemination of packaging knowledge. Centralized organizations usually have clearer structures to disseminate knowledge in a unified way to all packaging development teams. The dissemination can involve formal training and the distribution of new packaging knowledge within the company, such as training about packaging material innovations and how to apply new packaging materials. This process should be acknowledged in order for a decentralized organization to disseminate packaging knowledge efficiently.

4 Packaging customization

Decentralization means the organization is quite small, responsive and has close links to local markets. This type of packaging development organization has sufficient capabilities to customize packaging based on customer requests. A centralized organization, on the other hand, is less flexible and generally less able to customize packaging development.

5 Internal competition in packaging development

In a decentralized structure, the different packaging development units may compete for packaging solutions, such as key performance indicators (KPIs) for quality cost, volume utilization, etc. Such competition can support organizational development, but unsound competition may lead to unwillingness to share knowledge. In centralized organizations, there is a lack of competition, which should be acknowledged in order to avoid inactivity in packaging development.

Function- and project-based organization for packaging development

A function-based packaging development organization uses a hierarchical structure where all the expertise, competence and staff related to packaging are gathered in a specialist function led by a packaging function manager. In contrast, a project organization consists of a cross-functional team led by a project manager. The functions represented depend on the project task. For clarification, the differences between project-based organizations and decentralization are shown in Figure 10.1.

As with centralization and decentralization, the characteristics and associated pros and cons of function-based and project-based packaging development organizations are discussed below to highlight the embedded trade-offs and areas to consider when choosing the organizational type. The packaging development characteristics, which depend on function- and project-based organizations, are sorted into the following five categories.

1 Approach to capture conflicting requirements on packaging

In a function-based organization, the packaging requirements are generally specified before packaging development begins. This makes it possible to make informed trade-offs between these requirements in a structured way. However, as the requirements are specified beforehand, or are fixed because the product is already developed, this presents a challenge to a functional organization when up-to-date information of stakeholder requirements on packaging has to be included. A related challenge is to maintain flexibility in the development process in order to adapt to a change in requirements. In a project-based organization, it is common to gather packaging requirements in an iterative process throughout the development process because the stakeholders, such as logistics and marketing, are involved in the project. Thus, contradictory and modified packaging requirements from different stakeholders are more likely to become visible and manageable throughout the development process. The total development time for product and packaging is usually faster in the project-based organization, because of parallel development processes. In a function-based organization, packaging development often comes after product development, or at least starts later.

2 General and context-specific packaging knowledge

The structure of function-based organizations supports deep packaging knowledge, while a project-based structure supports more extensive context-specific packaging knowledge. Functions can gather and maintain in-depth expertise in different areas, for instance in packaging materials and technology. A general challenge is then to detect needs and opportunities to replace current knowledge with radical ideas, such as replacing wooden pallets with a new load carrier. In contrast to function-based structures, project-based organizations are more likely to develop applied packaging knowledge for a project, since it facilitates a holistic view of the goals and the context of the project. However, project organizations have fewer resources to develop in-depth packaging knowledge.

3 Inclusion of stakeholder requirements on packaging

In a function-based structure, there is a trade-off between standardized and flexible packaging solutions. It can be time-consuming to gather packaging requirements from various stakeholders. Even though all packaging requirements are supposed to be included beforehand in a function-based organization, there is a risk of missing some requirements from potential stakeholders. For instance, logistics may realize that the packaging solution did not consider requirements on stackability or volume efficiency. In a project organization, stakeholders work together on a project. Accordingly, this supports the inclusion of the requirements from the multidisciplinary project team. However, not all potential stakeholders may be included in the project team. Even if customers and suppliers are not included, for instance, their packaging requirements should still be captured in some way.

4 Packaging authority in the organization

Without packaging authority, there is a risk that packaging will be given less priority compared to other organizational considerations. This may reduce supply chain effectiveness and efficiency. In a purely project-based organization with a decentralized structure, it can be a challenge to maintain packaging authority because the packaging resources are divided among many sub-units. A similar challenge can occur in centralized project-based organizations, but here the centralized partner can be a stronger spokesperson for packaging. The function-based organization can be described in a similar way. If it is combined with *decentralization*, all packaging competence in the *sub-unit* is coordinated. If it is combined with *centralization*, the packaging competence in the *company* is coordinated. The greater the coordination of packaging resources, the stronger the authority it is likely to provide.

5 Focus on packaging or the packed product

The choice between a function-based and a project-based organization leads to a trade-off between packaging authority (and packaging competence) and supporting a focus on the packed product. A function-based organization focuses mainly on packaging and packaging development, while a project-based organization facilitates the focus on the packed product. By focusing on packaging, the function-based organization runs a risk of missing application-specific requirements on packaging, but it does have resources for in-depth packaging knowledge. By being involved in the development of both the product and the packaging in a project, such an organization offers a holistic approach to the packed product where the combined performance of packaging and products in the supply chain can be considered. The challenges are to maintain packaging authority and, for project participants from packaging, to focus on the overall supply chain performance rather than the packaging features.

Concurrent product and packaging development

The four generic structures for organizing packaging development support concurrent product and packaging development to various degrees, but it is possible to achieve it in all structures. Concurrent product and packaging development has several benefits. By developing the packaging system in the same phase as the product, it is possible to consider interactions and adapt the product and the packaging system to each other. This increases the likelihood of having packed products that are efficient from logistics, transportation and marketing perspectives. An example of this is a shelf-ready secondary package that is volume efficient in transportation and fits perfectly on a supermarket shelf. Another example is products in some primary packaging, such as soft drink cans, that can carry much of the load without secondary packaging.

The concurrent product and packaging development approach enables efficient packed product solutions. It should be noted that the packed product should embrace the functional requirements from end to end of the packed product including all stages from material sourcing, via packaging usage, to waste management. If the packed product is sold in different sales channels and different countries, the functional requirements often vary. To capture such differences, requirements mapping is needed.

The development time for new products can be quite long – in some industries, several years. With concurrent product and packaging development, the total development time of the packed product can be reduced. This enables purchasing to fulfil material sourcing specifications, packaging engineers to design packages and select packaging material with care etc. Furthermore, instead of adding the packaging development time after the product development time, they are done in parallel. Another effect of the long product development time is that sales must be forecast with rough estimates of sales volume and sufficient apportionment of packaging. If the early forecasts differ considerably from reality, there may be a need to update the packaging solution.

Summary

This chapter began by emphasizing that the way packaging development is organized determines a company's ability to integrate various requirements into the packaging system solution. This in turn affects the performance of packaging in supply chains. The chapter highlighted how supply chain performance is affected by an integrative approach to packaging development. Four generic organizational structures for packaging development were presented, and the characteristics, pros and cons of each structure were then discussed.

An integrative packaging development approach involves developing primary, secondary and tertiary packaging as a unitized packaging system that integrates packaging development with product development, logistics and other functional areas in a company, and involves customers and suppliers in the packaging development process.

The four generic structures for organizing packaging development are based on whether packaging development is centralized or decentralized in an organization, and whether it is organized in functions or as projects. In the first structure, packaging development is based on functional and centralized organizations. Packaging development is a support function that develops packaging systems for the sub-units based on their requirements. The second structure is functions-based and decentralized. Here, packaging is developed in sub-units without a centralized coordinator. The third structure is a project-based and centralized one. This means that packaging development is primarily carried out in a centralized support function, but there are packaging resources in the sub-units as well. The fourth structure is a project-based, decentralized structure, where all packaging development activities are conducted in the sub-units.

The chapter separately discussed the characteristics, pros and cons of centralized and decentralized packaging development organizations, and then of function- and project-based organizations. This accentuated the embedded trade-offs of the organizational choice. The characteristics of packaging development, which depend on centralization and decentralization, were sorted into five categories: 1) common packaging principles in a company; 2) management and division of packaging development resources; 3) capturing, maintaining and disseminating up-to-date packaging knowledge; 4) packaging customization; and 5) internal competition in packaging development.

The packaging development characteristics that depend on function- and project-based organizations were sorted into five categories: 1) approach to capture conflicting requirements on packaging; 2) general and context-specific packaging knowledge; 3) inclusion of stakeholder requirements on packaging; 4) packaging authority in the organization; and 5) focus on packaging or the packed product.

The chapter ended by highlighting the benefits of concurrent product and packaging design, which is more or less achievable in different organizational structures. The main message was that this increases the likelihood of having packed products that are efficient from logistics, transportation and marketing perspectives.

Packaging innovation capability 11

As discussed earlier in this book, there is great potential to improve both the effectiveness and efficiency of supply chains by making informed decisions about packaging and addressing packaging logistics in a novel way. The extent to which this potential is exploited is affected by an organization's innovation capability for packaging. In general, this innovation capability depends on how a number of processes support and enable innovation; in particular, the processes for packaging concept development and packaging development. Packaging concept development is the identification of new packaging concepts, while packaging development is the application of a general packaging concept to develop specific packaging solutions for given products. In addition, packaging innovation capabilities depend, to some extent, on the organization's views and practices in process innovation and technology acquisition, but these areas are not discussed here. The implementation and exploitation issues of packaging innovation are also outside the scope of this book.

Packaging concept innovation processes

As with all types of innovation, there are differences in packaging concept innovation. A well-known general difference is the one between incremental and radical innovation. Incremental innovations refer to minor changes where the general behaviour and knowledge are maintained and built upon, whereas radical innovations lead to fundamental changes in activities, behaviour and knowledge. A radical innovation may destroy the value of previous knowledge. In addition to this distinction, there are other packaging concept innovations that need other categories that can appropriately describe them.

One way to innovate packaging concepts is to develop concepts based on new knowledge about new materials or technologies for packaging or new

Figure 11.1 Typology for four generic packaging innovation concepts

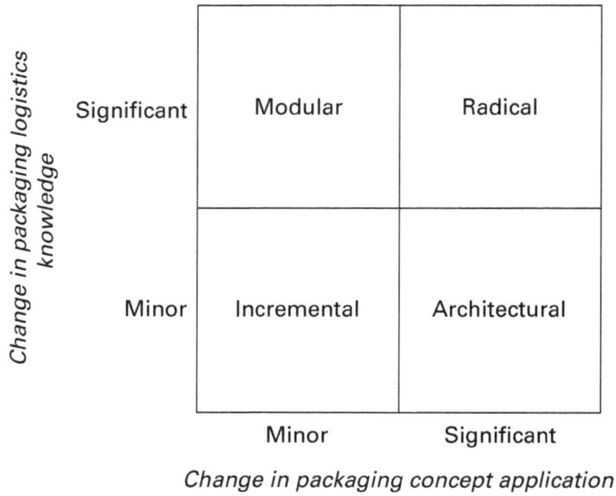

SOURCE Inspired by Henderson and Clark, 1990

equipment and processes in logistics. Another way is to make changes in the application of packaging concepts. Using these two distinct categories, the typology in Figure 11.1 defines four generic packaging innovation concepts.

The boundaries between the four types of packaging innovation concepts are not absolute, but the model can help to describe the type of innovations. This, in turn, can help to generate different kinds of packaging concepts, and support different perspectives in the process for packaging concept innovation.

The four types of concepts are incremental, modular, architectural and radical. *Incremental* packaging concept innovation refers to incremental improvements in a primary, secondary or tertiary packaging, which usually result in some functional improvements to a packaging system used within an existing packaging concept. *Modular* packaging concept innovation is based on new materials or technologies for packaging or new logistics equipment, which can result in significant changes in packaging functions or system design to be used within an existing packaging concept. *Architectural* packaging concept innovation is applicable when new market opportunities for packed products arise. Contextual modifications of current packaging concepts can then be developed for that market. Finally, *radical* packaging concept innovation is valid when both new market opportunities and new materials or technologies for packaging, or new logistics equipment occur. Major changes in how packaging concepts are both used and designed can then be developed.

Packaging concept development

To identify new concepts for primary, secondary and tertiary packaging, companies need to receive input on new research and development, new requirements (eg due to legislation and changes in the supply chain or in the market), new technological opportunities, new materials, new design ideas and new applications from various sources. Companies should then create, test and evaluate draft packaging concepts in order to identify those that are cost-efficient and feasible. To support this process, Figure 11.2 presents a stepwise structure for developing packaging concepts.

Information and idea generation

Information and up-to-date knowledge are key elements for successful innovation processes. Companies need access and establish linkages to various sources of information and knowledge to maximize the potential for packaging concept innovation. Organizations must monitor market needs systematically. To do so, they should have organizational structures that support regular communication with customers, in particular, long-term and close collaboration with their most important customers. Organizations

Figure 11.2 Packaging concept generation process

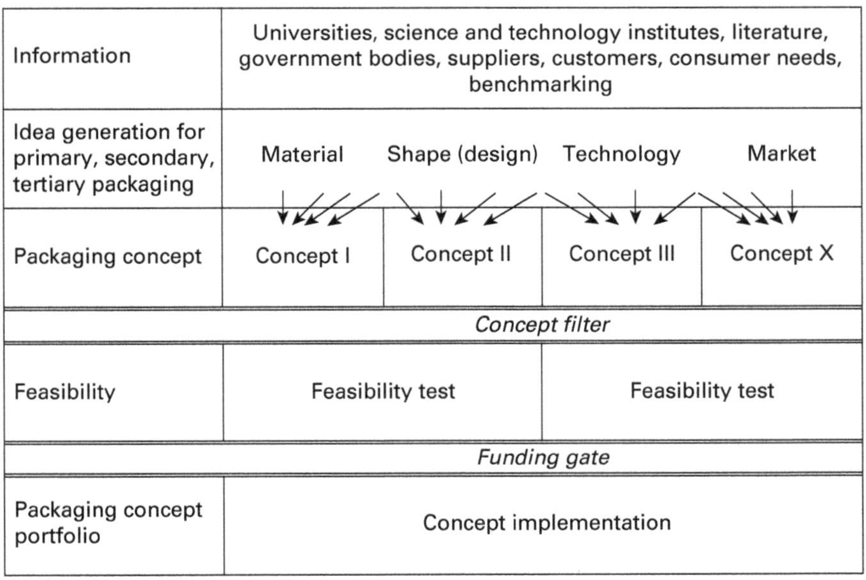

SOURCE Based on du Preez, Louw and Essmann, 2009

also need to screen, both internally and externally, for new packaging concept ideas and opportunities. However, securing customer feedback and capturing internal and external ideas are not sufficient. Companies also need to be able to analyse and use the information in their packaging concept development.

The information should be captured from various sources. Universities and journals provide scientific insights and news on the latest discoveries in packaging material, logistics and technologies, for instance. Science and technology institutes have insights in similar areas, often with a time lag, and with a focus on applied knowledge of new technologies and new packaging materials. Thus, it is essential that the corporate network includes contacts with universities and institutes in order to access these kinds of information. In terms of literature, companies should subscribe to relevant scientific and applied journals as well as industry-specific publications. Governments and government bodies develop new legislation and regulations related to changes in the type and usage of materials, newly discovered hazardous substances, material combinations, nutritional information, traceability requirements, recycling considerations and various text labelling requirements. Similarly, packaging suppliers' knowledge should be captured to facilitate new applied packaging innovation concepts. The packaging requirements and ideas from product or raw material suppliers, customers as well as consumers should also be embedded in the packaging innovation process. In so doing, external perspectives are added that can help new ideas grow. Further insights into the innovation process can be obtained through benchmarking: new ideas can come from scrutinizing competitors' packages and packaging concepts, and from the concepts of companies in other industries.

The information and the ideas from all these sources may refer to the packaging material, to the shape or the design of packaging, to the technology for producing, designing, measuring or tracking packaging systems, or to the market potential. Each of these areas may refer to a new or a differently used packaging material, technology, etc. For example, a well-known material can be applied in a new context or combined with other materials in an innovative manner.

Idea generation can also be about changing the shape or size of the packaging system. Such ideas should analyse the trade-offs in the packaging system and in the supply chain. A need for change should be the starting point for ideas related to the shape or the design of packaging. This need can be to have a more attractive design from a selling perspective, to increase the volume or weight efficiency, to reduce the amount of packaging material, etc. Having determined the need, the idea may be to change the shape or design of the primary, secondary or tertiary packaging in order to stand out in the sales channel against competitive products. It can also be to develop more volume- and/or weight-efficient packaging systems. Here, it is particularly important to consider the whole packaging system and the interactions between the various levels. For instance, the idea can be to reduce the need for secondary packaging by strengthening the primary or tertiary packaging. Another can be to reduce empty space in the secondary or the primary packaging in order to utilize the primary packaging or even the product itself to support the system's stackability. This can increase the total packaging system strength considerably, which means a reduced need for material. The following example comparing the strength of unfilled and filled secondary packaging of corrugated board illustrates the potential of such ideas. The primary packaging was a cardboard box with napkins. For one product, a box compression test (BCT) showed that support from the primary package increased the strength of the packaging system approximately three times, while another weaker primary package gave no support at all (Suhaimi, 2017, p 49).

Ideas for technology enhancement in packaging production, packaging design practices, measurements or tracking capabilities cover a wide and rapidly progressing area. Ideas related to packaging production should improve production and logistics efficiencies or produce more customized or product-oriented packages.

Technologies that enable efficient production can relate to improved filling efficiency, new equipment for automatic handling, or the inclusion of labelling in the packaging production process that does not reduce the production speed. An example of a technology that supports customized packaging is to have equipment with the flexibility to produce packages in various dimensions with minimal set-up times. The customized packaging solutions from such equipment enable the production of packaging on demand when needed. This facilitates the ability to minimize the amounts of air and filling material in the packages.

The starting point for coming up with ideas to modify packaging design practices should be the mapping of current practices. It may be that design practices already involve cross-functional teams with representatives or regular input from logistics, marketing, environment, production, transportation and purchasing; if not, then improvements can be made. Ideas can be generated for ways in which to set up these teams. One particular consideration is whether the company uses concurrent product and packaging design. This is primarily an organizational idea, but as discussed in Chapter 10, this can enable an effective and efficient packaging system where informed decisions regarding trade-offs have been made. The next logical consideration is the extent to which packaging requirements from suppliers, customers and consumers are included and captured in a structured way in the current practices. The information can be captured in various ways, such as through focus groups, market surveys, interviews, oral communication, workshops and archival data. Archival data, for instance, can show historical data of cubic utilization in transportation and product waste from inadequate packaging. This can provide input for the redesign of packaging or be included in the practices for packaging solutions for new products by considering historical data for products with similar features. To support new packaging design practices, companies should consider software and other technical support for pallet configuration, for determining the environmental impact of a packaging solution and for benchmarking the packaging design options. Examples of such tools are Cape Pack, PIQET, PackageSmart, EPIC, COMPASS and various LCA and simulation tools. In addition, some companies use internally developed tools.

Ideas for technology development and application related to measurements can involve new opportunities to capture new types of data, such as having sensors to monitor and record humidity, temperature and vibrations in supply chains. New technology may also enable advanced measurements for volume and weight utilization. It is often beneficial to combine volumetric and weight measures for packaging systems and transportation, but it may be complicated to capture the data needed for advanced measures. It is quite uncommon in current practices to measure the volume utilization of the packaging system. Finally, the technology enhancement of ICT can improve the tracking capabilities of packaging. Opportunities include capturing real-time tracking and traceability data with various identification technologies, such as RFID technology, near-field communication (NFC), quick response (QR) codes, or barcodes. In addition, it should be noted that as new technologies with new possibilities are continuously being developed, the costs for current and well-known technologies are changing. Thus, as costs go down, new possibilities may open up for a well-known, established technology. For detailed information about ICT in packaging, see Chapter 6.

Finally, ideas can be generated to use packaging to expand market share. Packaging concepts can be developed for new markets, or current concepts can be applied in new markets. In such developments, contextual insights from customers and consumers are valuable. By gaining insights into market and supply chain characteristics, packaging concepts for new markets can promote sales through consumer insights and achieve supply chain efficiency. The packaging concepts can thereby fulfil requirements in the best possible way. Another possibility is to develop new packaging concepts for the current markets. By reviewing the company's current packaging concepts and comparing them to supply chain and market requirements, and benchmarking them against competitors' packaging concepts, new or modified concepts can be developed that differentiate and fulfil requirements.

Packaging concept

After generating several ideas, the next step is to create feasible packaging concepts from one or a combination of ideas. Several concepts should be generated. To increase the likelihood of creating high-quality concepts, the development should be conducted by cross-functional teams with clear leadership and decision-making authority. These teams should work in well-defined processes for guidance in the concept development step. The combination of ideas into a packaging concept should also be carried out in parallel but integrated activities. This means that if two ideas are combined into a packaging concept, the complete concept (ie the combined and integrated ideas) is discussed and adjusted by the entire team, while the detailed development and application of the two ideas are done in parallel activities by smaller groups. In this way, the development process for packaging concepts can be both effective – with well-developed and integrated ideas – and time efficient. After finalizing a draft concept, it should be shared and discussed with a broader audience in the company. Its strengths and weaknesses can thus be further emphasized and modified if needed. There are various ways of doing this, such as in a workshop.

The outcome of the concept development process depends on the innovativeness and creativity of the organization. To support this, it is vital to have appropriate human resources in critical roles for innovation and to reward entrepreneurial and creative thinking. It is also essential to have an organizational culture that promotes and supports initiatives and ideas for new packaging solutions from employees, which can lead to unplanned packaging initiatives. To develop or scrutinize initial packaging initiatives and ideas for comprehensive packaging solutions, it is helpful if the organizational structure facilitates circulation among a cross-functional group for instant feedback.

Several features are essential for the concept development process to function over time. The processes for planning innovation, concept generation and implementation of new packaging concepts need to be integrated in order to secure a common direction. In addition, a key to long-term packaging solutions is to align the packaging innovation plan with the corporate strategy. By so doing, packaging concepts are developed to support the overall direction of the company. For instance, a company with a low-cost strategy would prioritize cost-efficient packaging solutions with high transport efficiency and inexpensive packaging material, whereas a company with a focus on premium quality is likely to prioritize premium packaging. A company with an extensive product portfolio may value brand recognition, but still have differentiated packages. A company with an environmental profile is likely to aim for packaging which is considered environmentally friendly.

Concept filter and feasibility

In the previous step, several draft packaging concepts were developed. This step now evaluates the feasibility of these drafts. The feasibility test includes analyses, calculations, modelling and prototyping. If needed, the evaluation process starts by collecting supplementary information. Based on the insights from the feasibility test, including new information compared to the draft packaging concept development, iterative changes to the draft concept are often necessary. The feasibility of the final versions of each packaging concept are then compared, from both a functional and a costing perspective. This final evaluation, also referred to as the 'funding gate', requires a business case analysis to prioritize among the packaging concept development projects. After that, decisions are made about which packaging concepts to fund for further development from a prototype to a final applicable packaging concept.

Packaging portfolio

The final step is the implementation of the funded packaging concepts. The packaging portfolio typically consists of a number of comprehensive packaging concepts that have several unique packaging solutions. This means that each packaging solution in a packaging concept shares a number of common attributes with the others. The packaging concept can be viewed as a platform that determines certain attributes, such as the graphic design, material or material performance, shape and labelling style, but does not have to include all of these. Other attributes of the individual packages in a concept are unique, such as dimensions, the material thickness and, if not included in the concept, the shape. The application of a packaging concept can vary. For instance, a packaging concept can be developed for a product category or for certain sizes of products.

The steps elaborated on in this section describe the development process for packaging concepts. This is one of a number of processes, which determine a company's packaging innovation capability. The next section discusses this capability.

Quality of the packaging innovation capability

The quality of the packaging innovation capability depends on seven processes. These were compiled by Chiesa, Coughlan and Voss (1994) as general innovation processes, but here they are contextualized to packaging. The quality of the first process, *packaging concept development*, already described in detail in the previous section, is determined by the extent to which various functions and external actors are involved, and the sources of information used. A high-quality process for packaging concept development has clear links to customers and suppliers, and captures consumer insights (if applicable) and supply chain requirements. It uses cross-functional teams with a broad range of competences in packaging concept development. A low-quality process is characterized by unplanned activities for developing packaging concepts, whereas a slightly more advanced process develops packaging concepts within a single packaging function. Here, it is common to have some customer contact. This approach can be advanced to capture detailed information in order to create new ideas for packaging concepts, often with a focus on the market and customers, but not to the extent of the high-quality process.

The second process is *packaging innovation planning*. The quality of this process depends on the time horizon considered for new generations or concepts of packaging. The most extensive planning processes consider three or more generations of packaging with a time horizon of approximately 5 to 10 years. This helps to establish a clear path to follow and to maintain logical linkages between packaging concepts over time. Slightly less comprehensive planning processes have two generations in the loop, or only one. Finally, some organizations do not plan at all. This indicates a low level of packaging innovation capability. Unless the most extensive planning process is implemented, it is recommended that a company add at least one packaging generation when planning its time horizon.

In order to be innovative, the quality of the third process, *inventiveness and creativity*, is essential. A company achieves high quality in this process by stimulating and rewarding entrepreneurial behaviour and innovative ideas. A practical consequence is being able to fund and support unforeseen activities. The extent to which companies build quality into this process is often done in stages. Before companies achieve this state-of-the-art process, they are likely to encourage risk taking and support new ideas, but are not fully prepared to reward entrepreneurs and fund unforeseen activities; they may also encourage new ideas, but avoid risk taking. Thus, the level of inventiveness and creativity is limited. The least inventive and creative companies focus on control systems with limited support for creative behaviour.

The fourth process, *packaging development*, describes the activities and procedures for developing packaging systems. It also describes the time horizon and the flexibility of packaging planning. In many companies, packaging has had low value and low status. It has been developed and added in at the end of the packaging development cycle. In this case, companies do not have established packaging development procedures, but rather use an ad hoc approach to developing or selecting packaging. There is great potential here for improving supply chain effectiveness and efficiency. A starting point to achieve quality in the packaging development process is to establish simple procedures that can be applied to all packaging projects. In the initial phase, though, these procedures are often sequentially applied without having parallel activities. This means that the process becomes inflexible and lacks time efficiency. In contrast, a high-quality packaging development process is clearly defined and has parallel and integrated activities. The objectives of the process are to be flexible in order to facilitate unexpected, time-dependent and small packaging projects so that they can be finalized rapidly. Before achieving this state of the packaging development process, companies tend to plan and review major packaging development projects in a structured manner. This usually generates qualitative packaging solutions in a fairly time-efficient manner for major packaging projects, but could be improved from a flexibility perspective.

Since packaging has multifunctional purposes and interacts with many parts of an organization, the fifth process, *teamwork and organization*, also needs to be evaluated to determine the quality of the innovation capability of packaging development. A high-quality teamwork and organization process should consist of multifunctional teams where representatives from logistics, marketing, sourcing, product design, transportation, environment, etc are involved early in packaging development. To make such a diverse team function efficiently with purpose-driven decisions, it is essential to have strong leadership. The team leader must have the authority to make decisions. A somewhat less qualitative, but still fairly effective, way to set up a team and organize packaging innovation is to include input from multi-functional teams before starting a packaging project, but there would be limited involvement from various functions in the development process. In such a set-up, there is clear project authority. A less qualitative approach is to ask for input from various functions to determine a packaging requirement specification before starting a packaging innovation project, but with limited interaction between functions before and during the project. This lowest-quality approach is characterized by having no teamwork and minimal inter-functional communication.

For a packaging innovation to become usable, it must fulfil or be adapted to *industrial needs and requirements*, the sixth process. This means that new packaging solutions should be resource efficient in the production and usage phases. For instance, the design of packaging should eliminate waste of time, material, energy and other resources that do not generate value. The most effective approach to secure the incorporation of industrial needs and requirements is to involve industrial designers as a key resource early in the packaging concept development process. A somewhat less, but still fairly, effective approach is to consult industrial designers a number of times during the packaging innovation process. These industrial designers can be either employed internally or hired in as external consultants. The earlier industrial designers are introduced into the packaging innovation process, the more likely the results will consistently be of high quality. Companies that do not consider industrial design clearly risk developing packaging solutions that are not adapted to industrial needs.

The final consideration in the seventh process of the packaging innovation capability is a physical realization of the industrial design. This means designing packaging solutions that can be *transferred to manufacturing and distribution* effectively and efficiently to function on a large scale. In addition to including the functions in the team discussed above, the company needs to ensure that the new packaging designs can be transferred to manufacturing and distribution by testing prototypes and ramping-up manufacturing of the new packaging solutions. It is also essential to manage engineering changes in manufacturing. A step towards such a high-quality process of transforming designs to manufacturing and distribution is to have clear links between manufacturing, logistics and packaging design. These links are determined in the teamwork and organization process. Companies that lack an organizational structure for this transfer process forward the finalized packaging design to manufacturing without interaction between these two groups. It

Table 11.1 Indicators for companies with high-quality processes within packaging innovation

Process	High-quality process indicator
Packaging concept development	Clear links to customers and suppliers Captures supply chain requirements and consumer insights Cross-functional teams
Packaging innovation planning	Consider three or more generations of packaging with a time horizon of five to ten years
Inventiveness and creativity	Promote entrepreneurial behaviour and innovative ideas Support and fund unforeseen activities
Packaging development	Clearly defined process with parallel and integrated activities Flexible activities and objectives to facilitate unexpected, time-dependent and small packaging projects to be finalized rapidly
Team work and organization	Multifunctional teams with strong leadership to be involved early in the packaging development process
Adaptation to industrial needs and requirements	Involvement of industrial designers as key resources already in the packaging concept development
Transfer new packaging designs to manufacturing and distribution	Include manufacturing and distribution functions in the packaging innovation team Test prototypes and ramp-up manufacturing of new packaging solutions Manage engineering changes in manufacturing

leaves the practical considerations, economies of scale, set-up times and other engineering concerns to be overseen by manufacturing, which can lead to inefficient manufacturing and time delays.

The seven processes together define the quality of the packaging innovation capability. They are summarized in Table 11.1, which lists indicators to achieve high quality in each process.

Summary

This chapter presented and elaborated on the packaging innovation capability, as well as on how a number of processes support and enable innovation. It distinguished between packaging concept development (identification of new packaging concepts) and packaging development (application of a general packaging concept to develop specific packaging solutions for given products).

A typology for four generic packaging innovation concepts was proposed that can help to sort innovations. This, in turn, can assist in generating different kinds of packaging concepts and can support different perspectives in the process for packaging concept innovation.

The chapter went on to propose a stepwise structure for developing packaging concepts. The first step is about information and idea generation. Companies need access and secure linkages to various sources of information and knowledge to maximize the potential for packaging concept innovation. A number of idea generation possibilities were covered. After generating several ideas, the next step is to create several feasible draft packaging concepts from one or a combination of ideas. The outcome of this process depends on the innovativeness and creativity of the organization, and a number of features for the packaging concept development process that need to be covered for it to function over time were discussed. The next step evaluates the feasibility of the packaging concepts. The feasibility test includes analyses, calculations, modelling and prototyping. This is followed by a funding gate. The final step is to implement the funded packaging concepts. The complete packaging portfolio typically consists of a number of comprehensive packaging concepts that have several unique packaging solutions.

The chapter ended with a discussion of seven processes that together define the quality of the packaging innovation capability. First, a high-quality process for packaging concept development has clear links to customers and suppliers, and captures consumer insights (if applicable) and supply chain requirements. It uses cross-functional teams with a broad range of competences in packaging concept development. The quality of the second process, packaging innovation planning, depends on the time horizon considered for new generations or concepts of packaging. The quality of the third process, inventiveness and creativity, depends on the extent to which companies stimulate and reward entrepreneurial behaviour and innovative ideas. The quality of the fourth process, packaging development, depends on the activities and procedures for developing packaging systems, such as the time horizon and the flexibility of packaging planning. The quality of the fifth process, teamwork and organization, is affected by the extent to which companies integrate multifunctional purposes and interact with many parts of an organization. The quality of the sixth process, industrial needs and requirements, depends on how it fulfils or is adapted to industrial needs and requirements. Finally, the quality of the seventh process of the packaging innovation capability is judged by its ability to transfer a packaging solution to manufacturing and distribution to function on a large scale.

12 Managing packaging logistics in e-commerce channels

Online retailing is growing rapidly in all the main Western markets. In recent years, the annual growth rate of e-commerce turnover in Europe has been about 15 per cent. The supply chain characteristics and requirements for e-commerce packaging are different to those for conventional trade. Supplying and distributing packed products in an e-commerce supply chain alters the cost and environmental impacts of the packaging material. In addition to the impact on the packaging material, e-commerce alters the cost and environmental impacts on five other factors: buildings, freight transport, passenger transport, unsold products and product returns (Pålsson *et al*, 2017), all of which except passenger transport are affected by packaging.

From an economic perspective, this chapter focuses on the changes in costs. However, e-commerce also has an impact on features related to sales and service, which in turn affect the costs and environmental impacts of the above five factors. For instance, many e-commerce companies tend to offer fast deliveries. To reduce the lead-time, some companies eliminate nodes between production and consumers. This increases the number of individual shipments and reduces the possibility to fill the truck. In order to facilitate fast response with efficient material handling and minimized stock-keeping costs of empty packaging, a limited number of packaging sizes are used. Thus, material-handling efficiency and stock-keeping costs are traded off against transport efficiency.

Cost and environmental impacts of packaging in e-commerce

The total amount and type of packaging material may vary between sales channels. For instance, clothing is usually sold in a plastic bag in stores, but requires more protection from corrugated board packaging or durable plastic packaging in e-commerce. Books are sold without packaging in stores, but are often individually packed in corrugated board in e-commerce. The types of buildings used in e-commerce are different from conventional trade buildings. In principle, e-commerce deliveries are sent from warehouses to pick-up points or consumers' homes instead of via a retail store (e-commerce products can also be shipped from a retail store, but for the purposes of this discussion, the more common pattern is used as the model). By avoiding stores, e-commerce reduces the cost and environmental impacts of the factor 'buildings', because warehouses usually consume less energy than retail stores. Thus, it is possible to design packaging for space minimization and handling efficiency in warehouses without having to consider potential trade-offs from packaging requirements in stores. Moreover, in the last mile delivery, e-commerce channels can use either pick-up points or home delivery. In home deliveries, freight transport to the home replaces passenger transport to a store. If the passenger transport is by car, the home delivery alternative reduces the cost and environmental impacts. From a packaging perspective, the cubic utilization of transportation from volume- and weight-efficient packaging is more important in e-commerce. Regarding unsold products, an e-commerce channel has greater opportunities to use a centralized warehouse, whereas each store keeps stock in the conventional sales channel. If each warehouse and each store have safety stock, the risk of accumulating unsold products because of obsolescence is higher in the conventional store-based sales channel. Finally, product returns are greater in e-commerce, particularly for experience goods, such as clothing. Thus, packaging should support returns because these products need to be repacked, which adds packaging material.

A number of examples illustrate the potential effect of packaging on the environmental impact of e-commerce. Packaging can determine whether home delivery and pick-up point systems are more or less efficient than a conventional delivery system via a retail store. In a study of DVD rentals, the packaging material was responsible for 67 per cent of the difference in energy consumption between the home delivery and the conventional distribution systems in favour of the store (Sivaraman *et al*, 2007). In another study of electronic products, the packaging material consumed approximately 22 per cent of the energy in the e-commerce distribution channel, and had a considerable negative impact on environmental efficiency (Pålsson *et al*, 2017). A third example, books in urban areas, showed that that e-commerce consumed significantly more energy per book than the conventional store because of more packaging material in the home delivery channel (Pålsson *et al*, 2017).

Even though packaging in many cases has major impacts on cost and environmental efficiency in e-commerce, current practices and research are quite undeveloped in this area. Current packaging solutions in e-commerce are often based on those originally developed for traditional commerce, but the supply chain characteristics and consumer demands on packaging are different in e-commerce. It is common to add secondary packaging to e-commerce products for protection during transport. There can also be challenges with the primary package. For example, blister packages can be used in a store to protect from theft, but when used in an e-commerce channel, an opened blister package leads to added costs and environmental impacts in product returns and repacking. As described above, supply chains with home delivery or pick-up points differ from those with a physical store, and so do the packaging requirements. Using packaging solutions developed for conventional supply chains (with a retail store) in e-commerce (with home delivery or pick-up points) may lead to unnecessary amounts of packaging material, low load factors and unnecessary product waste.

Delivery alternatives in e-commerce

E-commerce changes the distribution channels, which in turn changes the packaging requirements. There are several alternative ways to organize distribution channels for e-commerce. A conventional business-to-consumer (B2C) supply chain involves a number of nodes, usually raw material suppliers, a manufacturer, warehouses and retail stores. The nodes are connected by freight transport, and consumers transport products from the stores to their homes. Instead of having a retail store, an e-commerce supply chain delivers products directly from a warehouse or a production plant to the consumers' home or to a nearby pick-up point. An alternative is for online orders to be collected in a store. Each of these three delivery options can be organized in different ways. Figure 12.1 presents an overview of common delivery alternatives in e-commerce. Other alternatives, currently small-scale models, are being developed.

Figure 12.1 shows the nodes and distribution links from the point of collection by the customer. Before that point, the supply chain from the producer (A, B and C in Figure 12.1) can be structured in a number of ways. In A, the nodes from the manufacturer to a conventional store often include one or a few warehouses. A similar supply chain structure can be used in B and C, but since the number of pick-up points and homes is much higher than the number of stores, it becomes more challenging to achieve a high load factor in B and C compared to A. Furthermore, it is also increasingly common to deliver directly

Figure 12.1 Delivery alternatives in e-commerce

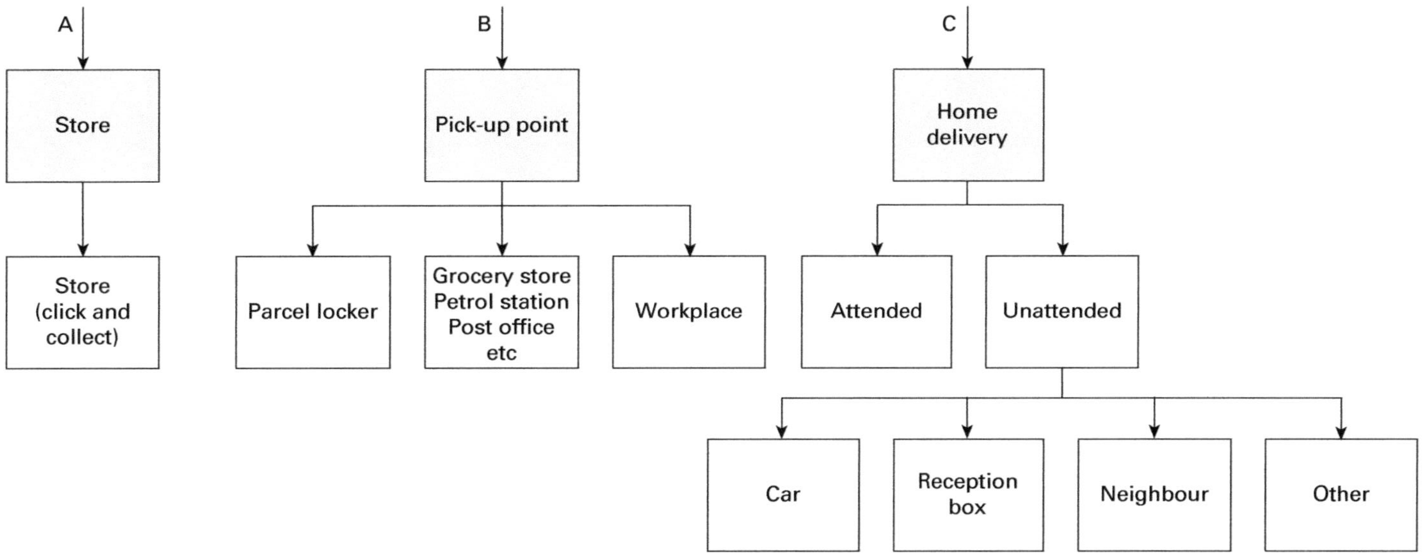

from the production plant to the pick-up point or to the consumer in B and C, which results in less co-packing and more frequent deliveries.

Click and collect means that consumers order online and collect the order in a store (this term can also be applied to other delivery models, such as parcel lockers). From a packaging perspective, this set-up is similar to a conventional channel. Pick-up points can be organized in one of three ways:

- A common option is for consumers to pick up the order at an assigned partner to the delivery provider. This could be different kinds of businesses, such as grocery stores, petrol stations or post offices.
- Another option is the utilization of parcel lockers, which can be placed at retailers, university campuses, companies or close to people's homes. Parcel lockers are self-governed reception boxes with locks, which can be used to receive or send parcels. The packaging dimensions cannot be greater than the dimensions of the reception boxes of the parcel locker.
- A third pick-up point option is workplace deliveries. These can be combined with a parcel locker, or delivered directly to the consumer at their workplace, or to a reception desk.

The implications for packaging with pick-up points are an increased focus on enabling transport efficiency, stackability, handling efficiency, and reduced need for having promotional attributes.

Home deliveries can be either attended or unattended. An attended home delivery means that the consumer receives the delivery. An unattended home delivery brings additional security considerations. These can include delivering to a neighbour, to a reception box with a lock, or in a smart car as a car drop, which means that a trusted delivery partner can open the trunk. Other options are being developed. In order for unattended home deliveries to function, protection against temperature and humidity may need to be increased. In addition, the packaging dimensions should match a reception box or a car trunk, and the track & trace capability of the packaging should help to confirm that a delivery has taken place.

Effects of e-commerce supply chains on the packaging system

As discussed in the introduction to this chapter, the cost and environmental impacts of B2C e-commerce and conventional trade channels depend on six factors in the supply chain: packaging material, buildings, freight transport, passenger transport, unsold products and product returns. Each of these factors can change as a result of changes in the supply chain. This section focuses on changes in the supply chain which impact on packaging. Even though the distribution channels related to e-commerce can be organized in different ways, a number of general packaging implications can be determined for e-commerce compared to conventional supply chains. The section describes changes in B2C e-commerce chains compared to conventional supply chains, and analyses the general implications on the packaging system.

The supply chain effects of e-commerce and their implications for packaging systems are illustrated in Figure 12.2 which combines an updated version of the implications of e-commerce on different parts of a supply chain discussed in Siikavirta *et al* (2003) with packaging features. The links to the six factors are shown in the figure.

Figure 12.2 shows that to obtain a cost and environmentally efficient packaging system, nine packaging features should be addressed differently for products in supply chains with home delivery or pick-up points than for products in retail store supply chains:

- product protection;
- apportionment;
- communication;
- volume and weight efficiency;
- packaging postponement;
- unitization;
- stackability;
- identification;
- packaging waste.

In sourcing, e-commerce with home delivery systems can utilize point-of-sale data to replace inventory with information for raw material supply, which can reduce overproduction and obsolescence of the product. Less product waste means less packaging waste as well. In production, e-commerce increases the demand for cost efficiency and short lead-times, which are easier to meet with pull-based production than with a push strategy. A pull-based system handles demand fluctuations better and thus reduces overproduction. A long-term trend in distribution is for e-commerce customers to demand faster deliveries with little or no delivery charge. To meet this demand, companies selling online need to have short delivery lead-times, which can affect the mode of transport, particularly for global companies

Figure 12.2 Supply chain effects of e-commerce and their general implications on the packaging system

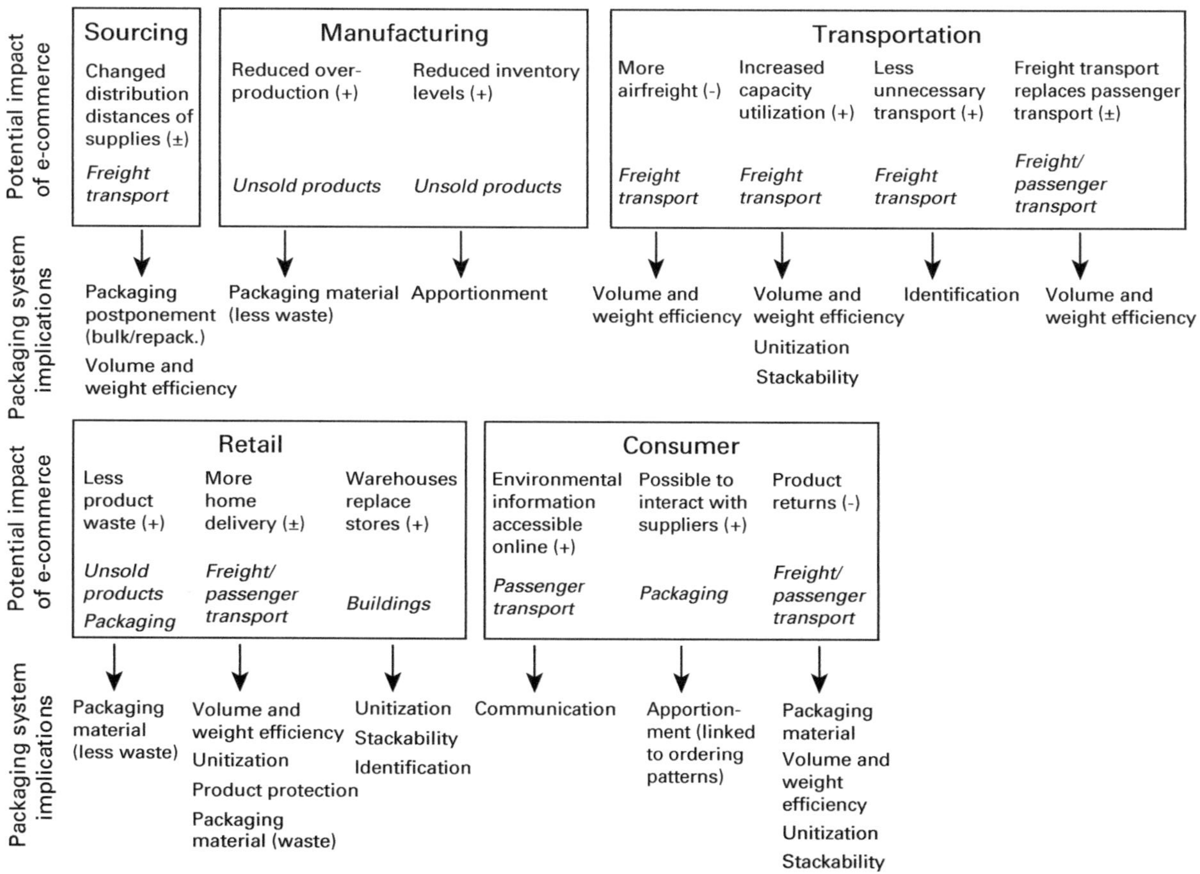

and for those with volatile demand. As a result, sea and rail transportation may be replaced by road and air freight, which both have negative effects on costs and the environment. In home delivery systems, products are delivered directly from a warehouse or a production plant. A warehouse costs less and consumes less energy per square metre than a retail store. Thus, if the products have the same turnover time, they cost less and consume less energy in buildings than those sold in conventional stores. The home delivery option also results in more packaging material ending up in the home instead of being removed in the retail store. In the consumption phase, e-commerce can enable new services, such as grocery replenishment to consumers, with more cost and environmentally efficient processes for the manufacturer. However, e-commerce also leads to more product returns, which means a return trip for the consumer, a return freight transport, repacking and administration.

Figure 12.2 highlights the effects on the nine packaging features in different parts of a supply chain in e-commerce channels. The effects are summarized in the following sections according to the packaging feature, and in Table 12.1.

Product protection

When products are sold in a retail store ownership changes in the store. This means that the responsibility for protecting the product shifts from the store owner to the consumer. In home deliveries and at pick-up points, the change of ownership is delayed. This means that the packaging needs to protect the product in another setting, such as a delivery truck, at home or at a pick-up point. Compared to a retail store, e-commerce packaging should enable additional stacking and handling activities. Another aspect of packaging protection is that in retail stores, temperature-sensitive products are protected by equipment that maintains the required product temperature, such as refrigerators, freezers and hot cabinets. Such equipment is less available in home delivery systems or at pick-up points. Thus, in these e-commerce delivery systems, it is beneficial if the packaging is able to protect against temperature fluctuations.

Apportionment

Packages should contain the right amount and have sufficient size in order to minimize the level of obsolescence and product waste. In supply

Table 12.1 Nine packaging features to be addressed differently in the development and selection of packaging in e-commerce vs conventional trade

Packaging feature	Consideration in e-commerce supply chains
Product protection	Enable additional stacking and handling activities Increased product protection against temperature
Apportionment	Reduced inventory levels in sourcing and production and elimination of adaption to retail shelves affect the right amount and size of the packaging system Increased focus on consumer need for the right amount and size of primary packaging Increased focus on picking efficiency in warehouses
Communication	The communication function of packaging can be modified by publishing some information on the web instead
Volume and weight efficiency	Increased use of airfreight leads to increased importance of volume and weight efficiency Increased importance of volume and weight efficiency for long-distance shipping and home deliveries
Packaging postponement	Greater need for bulk transportation with repacking due to consumer-packed products from the warehouse
Unitization	Unitization becomes more important to utilize transport capacity (longer distances, co-packed with competitors) and minimize storage space in warehouses
Stackability	Stackability becomes more important to utilize transport capacity (longer distances, co-packed with competitors) and minimize storage space in warehouses
Identification	Packaging labelling and identification should be more detailed, particularly on the primary packaging Identification and digitalization can make warehouse operations more efficient
Packaging waste	Less packaging material waste in production and retail stages More packaging material waste ends up at the consumer stage More packaging material from product returns

chains with home deliveries or pick-up points, total inventory levels in sourcing and production can be reduced, because of shorter lead-times and centralization. This affects what is determined to be the most efficient amount and size of a given packaging system. For instance, secondary packaging can be designed to match picking efficiency and warehouse requirements without making trade-offs for retail store requirements. Furthermore, because there are no retail stores in e-commerce supply chains, decisions about the dimensions of the primary packaging do not have to take into account retail shelf requirements in order to match consumption needs.

Communication

The e-commerce web page can provide more product information than stores usually do. That is where consumers can obtain additional environmental information in the process of deciding what to purchase. Similar information for products sold in stores often needs to be presented on the packaging. Accordingly, in supply chains with home delivery and pick-up points, the communication function of packaging can be modified.

Volume and weight efficiency

A supply chain with home delivery or pick-up points instead of a retail store has implications for how to address the volume and weight efficiency of the packaging system. In home delivery systems, passenger transport from the retail stores is replaced by freight transport to the homes. If organized accurately, this transport process can be carried out more energy efficiently by a transport provider than by a consumer. The efficiency depends on logistical prerequisites, such as the number of consumers, geographical distance, delivery times and more, but also on the volume and weight efficiency of the packaging system to maximize the cubic utilization of the vehicles. The importance of cubic utilization is also more important because of more product returns in e-commerce, which leads to more freight transport. Furthermore, in some e-commerce channels, long-distance shipping and use of air freight are more common because of central warehouses. For those products, the importance of optimizing the volume and weight efficiency of packaging increases. From a cost perspective, it is the weight efficiency in particular that should be reduced, as it usually determines the freight cost. In terms of greenhouse gas emissions, both volume and weight efficiencies are essential.

Packaging postponement (bulk/repacking)

The nodes and links in supply chains with stores are often different from those with home delivery or pick-up points, as described with the three arrows (A, B and C) in the text related to Figure 12.1. In e-commerce channels with centralized warehouses and long transport distances, the products are dedicated to a certain consumer earlier in the supply chain than in a store, thus moving from push to pull. In total, these differences affect the ability to apply packaging postponement. In global supply chains, for instance, bulk packaging with repacking that is carried out close to the consumers could improve transport efficiency, particularly since many different products should be co-packed in the final stage of home delivery and pick-up point systems to be environmentally efficient. In this case, the packages benefit from being standardized or modularized for stacking and handling efficiency. As a result, though, each package cannot be optimized according to volume efficiency. Thus, the final customized packaging should be transported as short a distance as possible.

Unitization

Unitization reduces the number of separate packages that need to be handled, thus minimizing the time and energy consumption in materials handling. Since packed products have a unique consumer earlier in supply chains with home delivery and pick-up points than with a retail store, the opportunity to obtain increased capacity utilization in distribution depends on the unitization possibility between various packed products. Thus, in supply chains with home delivery or a pick-up point, packaging unitization should consider both the secondary packaging of the particular product and the dimensions of other packed products, as these can be co-packed in the same transport. Unitization also becomes more important as warehouses replace retail stores in e-commerce and because product returns increase. To minimize storage space in warehouses and facilitate co-loading in return transport, packaging should be unitized.

Stackability

Stackability is related to unitization in the sense that the latter usually facilitates the former. However, stackability also depends on other issues, such as packaging strength and stability. Stackability is important for transport, handling and storage efficiencies. These three areas often have a great impact on e-commerce channels, which increases the importance of stackability in e-commerce.

Identification

For home delivery and pick-up point systems to be efficient, packaging labelling and identification should contain more detailed data compared to retail store chains. The primary package, in particular, needs to be linked to the consumer, which is often unnecessary in retail chains. In addition, e-commerce, in combination with other digitalization, can enable logistics structures from long-distance shipping with centralized warehouses to immediately sort products for the final customer destination after production. The argument for doing this is that warehouse operations – such as handling, sorting and picking – are currently often more costly than long-distance shipping (in developed countries). Although digitalization can empower cost-efficient sorting with a minimal need for manual labour, it requires packaging identification through labelling. Digitalization should be considered for all types of sales channels, but e-commerce offers the greatest savings potential, both economically and environmentally. An additional benefit of implementing identification technology through packaging labelling is reduced shrinkage and product waste.

Packaging waste

As a result of less product waste in the production and retail stages in e-commerce, there is also less packaging material waste in these stages. On the other hand, the packaging material, particularly from secondary packaging, ends up to a greater extent at the consumer stage rather than at the retailer. Current packaging solutions were often originally developed for conventional stores. The same package is often used in the e-commerce channels, with added packaging material if needed for product protection. Thus, unless a separate packaging solution is developed for the e-commerce channel, it ends up adding packaging material. In general, e-commerce also generates more packaging waste related to more product returns. Each product return is usually repacked into a new secondary package.

Summary

This chapter presented and discussed the economic and environmental impacts on packaging in e-commerce channels. It compared packaging in e-commerce to conventional supply chains with stores. It highlighted that supplying and distributing packed products in an e-commerce supply chain alters the cost and environmental impacts of six factors: packaging material, buildings, freight transport, passenger transport, unsold products and product returns. It also described how each factor except passenger transport is affected by packaging.

The total amount and type of packaging material may vary between e-commerce and conventional channels. This is illustrated by a number of examples. The current packaging solutions in e-commerce are often based on ones that were originally developed for traditional commerce. This is problematic because the supply chain characteristics and consumer needs for packaging are different in e-commerce.

The distribution channels related to e-commerce can be organized in different ways. This chapter presented an overview of common delivery options in e-commerce. Despite the different delivery alternatives, it was possible to identify a number of general packaging requirements that need to be emphasized differently in e-commerce compared to conventional supply chains. These packaging requirements were identified by mapping the supply chain effects of e-commerce, and then analysing the implications for the packaging system.

As a result, nine packaging features were identified that should be addressed differently for products in e-commerce supply chains: product protection, apportionment, communication, volume and weight efficiency, packaging postponement, stackability, unitization, identification and packaging waste. These packaging features and their general impacts were summarized in Table 12.1. These changes are general in nature and do not apply to specific cases. Many companies also utilize both conventional and e-commerce sales channels, which may affect their packaging solutions. From the economic and environmental perspectives, it is necessary to analyse carefully whether it is beneficial to develop dedicated packaging solutions for each sales channel.

Managing packaging logistics in developing countries 13

Packaging systems should be evaluated and designed using similar methods in both developed and developing countries. Although each country categorized as 'developing' has its own individual characteristics, certain supply chain challenges relating to packaging are common to many, and require special attention. The supply chains often include several stages where packages are filled and handled manually. These manual-handling processes are often rough, resulting in shocks to the packaging. It is less common to use pallets, which reduces unitization and handling efficiency, but is more common to stack higher than in developed countries, even though the conditions might be worse, such as humidity. The infrastructure is less developed and the road conditions are sometimes unsatisfactory.

Packaging features requiring special attention in developing countries

In general, supply chains in developing countries often have challenges related to the following categories (Parfitt, Barthel and Macnaughton, 2010; Saruchera, 2017; Souhrabpour, Hellström and Jahre, 2012):

- infrastructure: eg bumpy roads that generate vibration for goods loaded in trucks;
- organization and knowledge: eg many stages in the distribution chain, and temporary, unemployed workers who help to unload goods;

- logistics equipment: eg lack of equipment with temperature control and lack of automatic handling equipment (instead there is rough manual handling of packaging);
- costs: eg minimizing the packaging cost;
- climate: eg humidity and temperature challenges;
- security: eg risk of counterfeiting or theft.

Certainly, these challenges also exist in developed countries, and do not appear in all parts of developing countries, but are more common there.

Food packaging in developing countries fulfils an important role in supplying food to consumers. There is considerable post-retail food waste in developed countries, while such waste is lower in developing countries (Bond *et al*, 2013). Instead, food waste there occurs mainly in production, transport and processing because of unsatisfactory transport and storage technology, and the lack of knowledge and investment (Parfitt, Barthel and Macnaughton, 2010). Food packaging therefore has many requirements including protection against physical shock and contamination, and temperature control.

To address the supply chain challenges of packaging in developing countries, it is most important to map and identify supply chain requirements in order not to over- or underpack. As a starting point when designing packaging for developing countries, Table 13.1 highlights common challenges and their effects on packaging in developing countries.

Infrastructure

The infrastructure for transportation and storage is often inadequate in developing countries. This means that packaging systems are exposed to rougher conditions to a far greater extent. As a result, packaging needs to provide better protection and clearly communicate handling and stacking requirements as well as fragility concerns. In many areas in developing countries the state of the roads is poor, at least compared to developed countries, with challenging road conditions including bumps, and gravel roads with mud or rocks. This is why packed products in trucks are often exposed to vibration, shocks and impacts. As a result, road transport places greater demands on the protection function of packaging systems.

The conditions for railway transport are also often insufficient. This was exemplified in a survey of companies' concerns for rail transport in Namibia. Among the 12 most essential challenges were lack of intermodal services, lack of essential transportation equipment, non-procedural practices at

Table 13.1 Common challenges and their effects on packaging in developing countries

Category	Challenge	Effect on packaging	Packaging feature
Infrastructure	Poor road quality	Vibration, shocks	Protection
	Poor railway quality	Rough loading, time consuming	Protection, communication
	Poor quality of buildings	Insufficient temperature resistance	Protection, communication
	Insufficient recycling systems	Less need for recycling	Biodegradable packaging
Organization and knowledge	Low credit to logistics	Packaging low status	Logistics training, communication, information
	Lack of training	Rough material handling	Protection
	Temporary workers	Rough material handling	Protection
	Lack of standardization	Inefficient logistics	Unitization
	Low level of collaboration	Incomplete packaging requirements	Improve packaging development
	Many steps in the supply chain	Much handling	Map the supply chain, protection, handleability
Logistics equipment	Old trucks	Vibration, dirt	Protection
	Old material-handling equipment	Inefficient handling	Unitization
	Lack of handling equipment	Manual handling and packing	Enable manual handling (weight, handles etc), facilitate manual packing operations
	Filling efficiency	Manual filling process	Easy to pack manually
	Inadequate equipment	Infestation by vermin (eg rats)	Protection
	Lack of ICT	Lack of tracking on packaging	Human-readable text, illustrations
	Lack of pallets	Inefficient handling, less protection	Stackability, unitization, protection

(*continued*)

Table 13.1 *(Continued)*

Category	Challenge	Effect on packaging	Packaging feature
Costs	Inexpensive packaging material	Inadequate protection	Protection
	Minimize packaging material (1)	Large packages	Apportionment, enable manual handling
	Minimize packaging material (2)	Thin packaging	Protection
	Low labour costs	Manual handling	Enable manual handling (weight, handles etc)
	Short transport distance	Local packaging sourcing	Material quality properties
	Storage space utilization	High stacking capability	Stackability
Climate	Temperature*	Protect against temperature	Protection
	Humidity*	Protect against humidity	Protection
Security	Theft	Hard to open, hard to identify manually, traceability	Security, ICT
	Counterfeiting	Hard to open, hard to identify manually, traceability	Safety Security, ICT

*Lacks logistics equipment to protect against temperature and humidity the rail ports, poor rail-transport management skills, and theft or pilferage of goods (Saruchera, 2017). From a packaging perspective, the lack of transportation equipment, ad hoc practices and poor management result in rough handling of packages and time-consuming packaging activities. Thus, the ability of the packaging to provide protection is emphasized. To reduce time-consuming packaging activities, communication on the packaging using text and illustrations will present handling instructions more clearly.

In terms of infrastructure, buildings may be of insufficient quality. Warehouses or other buildings for storage may not have tight walls and may lack insulation. With poor building quality and insufficient temperature resistance, there are risks that packages may be too hot or too cold. As with the other infrastructure challenges, this means that protection, particularly for insulation, needs to be emphasized. Packaging communication can also facilitate correct storage, for instance, by having labels that illustrate the temperature restrictions.

Organization and knowledge

Organization and knowledge of packaging and logistics greatly affect logistics performance. However, in developing countries, logistics is generally given low credit (Kovács, and Spens, 2009). There is a lack of training programmes for logisticians and staff working with packaging. A similar observation was made regarding poor rail-transport management skills in the infrastructure category (Saruchera, 2017). Another observation in South Africa was that limited knowledge and economic constraints resulted in trucks that were overloaded beyond the legal limit, and in the use of lower order roads instead of national routes to avoid tollgates and weighbridges (van der Mescht, 2006). Because of these kinds of organizational and knowledge challenges, packaging logistics has a low status and due to the lack of training, material-handling skills are poor, often unnecessarily rough. The lack of trained logisticians together with limited organizational structures and pressure to use cheap labour result in the use of temporary workers. Local people, not associated with the company, may help to unload a truck or move packaging. This exacerbates rough material handling, as temporary workers not only lack training, but lack loyalty to the company.

The long-term solution to these challenges is to provide logistics training and avoid temporary unorganized workers by having control of the supply chain. However, these kinds of changes take time. Large companies are involved in numerous supply chains, each one containing many materials-handling situations. Thus, to securely keep products from being damaged, it is essential to analyse the appropriate protection needed for the packaging system. In many cases, additional protection is needed compared to similar supply chains in developed countries. Another complementary approach is to include labels with sufficient information about how to handle the packages. It is recommended that these labels include illustrative pictures as much as possible instead of text to overcome language barriers.

A number of challenges in the organization and knowledge of packaging and logistics are related to the lack of standardization in logistics. As outlined in Chapter 2, standardization aims to reduce complexity, target selected activities, protect the interest of majorities, persist over time, but be subject to review and change when needed. Logistics standardization can be sorted into packaging, transportation, material handling and ICT standardization. Developing countries usually have a low level of logistics and packaging standardization, which generally results in inefficient logistics.

With a low level of standardization in the dimensions of packaging and load carriers, it becomes a challenge to maximize cubic utilization of vehicles, co-load unitized packaging, and simplify material handling. Furthermore, a lack of standardized recycling guidelines for packaging waste reduces the levels of recycling. Limited transportation standardization in developing countries makes it a challenge to fit as many packed products as possible into vehicles. Without standardized vehicle dimensions, it is also more challenging to use intermodal transport, because of the lack of modularity. Limited material-handling standardization, such as storage and material-handling equipment, which should be aligned with that of packaging, also reduces resource efficiency in developing countries.

However, it is possible for authorities to support material-handling standardization. An approach to doing so in one developed country, South Korea, was described by Min *et al* (2014), and can provide inspiration. There the government agencies offer discounts on storage fees and favourable loans to users of standardized storage equipment. Another approach is to use a certification system for standardized logistics equipment where users receive benefits, such as tax reduction or subsidies.

The final logistics standardization dimension is that of information technology. In developing countries, ICT standardization as well as the general technology level are not as advanced as in developed countries. However, ICT standardization can enable supply chain actors to coordinate physical material and delivery processes. Thus, if financially possible, ICT standardization can provide substantial benefits in developing countries.

The category of organization and knowledge also encompasses collaboration, which in developing countries is at a low level. Supply chain collaboration is the starting point for the coordination of activities and processes within and between firms. This can help to avoid sub-optimization in supply chains, such as through the avoidance of unnecessary repacking or mismatched packaging and logistics equipment. Since packaging has an impact on all logistics activities in the supply chain, numerous requirements from supply chain actors must be considered in order to design or select appropriate packaging systems. For instance, requirements on protection, handling, promotional attributes, and storing from producers, warehouses, transport providers and retailers should be combined and analysed as a unit. As discussed in Chapter 6, the track & trace capabilities of packaging also facilitate supply chain communication. However, with a low level of collaboration, as in developing countries, it is more challenging to carry out the holistic analyses necessary for considering supply chain requirements in packaging design and selection, and to capture and utilize track & trace data for logistics efficiency. As a result, packaging systems are likely to consider incomplete requirements, resulting in missed opportunities for balancing costs and services throughout the supply chain.

A final challenge related to organization and knowledge is that supply chains in developing countries often have many stages (Parfitt, Barthel and Macnaughton, 2010). The distribution chains, in particular, often have several stages before the packed product reaches the consumer, meaning that packages are handled many times in different environments: repacking in a warehouse or outdoors, movement from one vehicle to another, carrying, throwing packaging, etc. This places material strength demands on packaging so that it can sufficiently protect the product. Mapping the supply chain can result in two options: 1) determine whether each stage is necessary, and if not, modify the supply chain; 2) for the remaining stages, the way packaging is handled can be mapped, and based on that, specify the handleability requirements that can facilitate the modification of the packaging systems used.

Logistics equipment

Packaging systems interact with logistics throughout supply chains. However, in developing countries logistics equipment is commonly inadequate. This presents a number of challenges that need to be addressed by packaging. Both trucks and material-handling equipment are commonly quite old and of lower quality (Soubrabpour, Hellström and Jahre, 2012). For this reason, packaging protection becomes more important in facilitating efficient handling, protection and more. Old material-handling equipment results in inefficient handling processes. An example is the use of overloaded vehicles. Unitization of packaging can increase the utilization efficiency of these vehicles.

The lack of handling equipment results in manual handling and packing processes. In fact, a majority of the filling processes are manual. An example from South Africa is shown in Figure 13.1. Manual handling also involves packaging carried in baskets and bags. To support this, the packaging should enable manual handling and facilitate manual packing operations, should be easy to carry, ie with handles, not be bulky, and have a suitable packaging weight. It should also be easy to open, close and fill.

There is often a lack of pallets in developing countries, leading to inefficient handling and less protection of the other packaging. To compensate for this shortage, it is essential to design packaging that can be stackable and unitized. By doing so, it may still be possible to handle, store and transport packaging efficiently.

Finally, logistics equipment and packaging lack advanced ICT labelling to a great extent, which means that it is difficult to track packages. To address this challenge, the packaging should have labels with both text and illustrations.

Figure 13.1 Outdoor packing facility in South Africa

Costs

Economic considerations are prioritized in developing countries: cost-efficient solutions are essential. This has at least three implications for packaging.

First, it is common to use inexpensive packaging material and to minimize the amount of material (Souhrabpour, Hellström and Jahre, 2012). This is problematic if it results in too little protection, which leads to product waste. One example is that a pallet of poor quality can bend, which results in compressed secondary packaging. Another example is inadequate packaging material which causes pallets and packaging to break because they cannot bear the weight. Still another example is nails on wood packaging such as pallets which can cut holes in other packaging (eg in secondary packaging). As discussed in Chapter 4, the total cost and environmental impacts are often higher if products rather than packaging material are wasted. In that case, it is often more cost efficient to use a rather more expensive material with more protective features or to add packaging material. To present rational arguments for such a course, companies need to have a systems approach towards packed products. This is challenging as collaboration in the supply chain is generally limited. Thus, there is an organizational and knowledge challenge related to this cost consideration. An intial step to explain the total cost perspective can be to make and present a total cost analysis of a packed product in a supply chain. The minimization of packaging material results in the use of large packaging (Souhrabpour, Hellström and Jahre, 2012), as this means that less packaging material is used per product. The objective is to minimize the cost of packaging material. In order for large packaging to work satisfactorily in the supply chain, one needs to consider apportionment and the enablement of manual handling (eg by including handles and avoiding shapes that are difficult to lift and move). Another aspect of the minimization of packaging material is to use packaging material that is too thin. This leads to inadequate protection with unnecessary product waste. The simplest solution is to add packaging material.

Second, the cost of labour is inexpensive compared to that in developed countries. As logistics equipment is expensive in comparison, manual handling is often the preferred option. The packaging features needed are those that enable effective and efficient manual handling including weight, volume and handles.

Third, to minimize transportation costs, a common practice in developing countries is to use local packaging sourcing (Souhrabpour, Hellström and Jahre, 2012). In this case, it is essential to secure the appropriate material quality. In order to avoid misunderstandings, it can be beneficial to describe and specify the properties and features of a material rather than specify the thickness or grading of a given packaging material. Certain differences in packaging material standards and interpretations of specifications can thus be avoided. Another practice related to minimizing transportation costs is to utilize storage space efficiently. Many packages are commonly stacked on top of each other. To facilitate this practice, packaging should have features that enable high stacking.

Climate

Packaging should protect against temperature and humidity. In many developing countries, these climatic aspects are challenging due to high temperatures and considerable humidity. Their effects may also be underpinned by inadequate logistics equipment. Warehouses in the form of tin sheds can become hot; other types of warehouses may not shut tight or may leak. Accordingly, for packaging to compensate for such challenges it needs to have extra protection features. Depending on the protection needed, these features may include providing insulation for temperature protection or withstanding moisture and humidity.

Security

Security in packed products involves two challenges: theft and counterfeiting. Because of poverty and a low standard of living combined with limited supply chain control, even inexpensive products are more likely to be stolen. Theft and pilferage are common in the railway sector (Saruchera, 2017). Counterfeiting is an increasing problem, the purpose of which is to sell cheap copies of more expensive products. This is a well-known problem, in particular for premium brands and pharmaceutical products.

Packaging can reduce both the risks of theft and counterfeiting by being hard to open, hard to identify visually and by having traceability capabilities. These features do not have to apply to all levels of the packaging system. If the tertiary packaging is anonymized, for example, it becomes more difficult to identify or detect a given product. In a similar manner, if the primary packaging is so well secured that it is obvious when it has been opened, it becomes more difficult to exchange products. Traceability capabilities can improve control of the supply chain and thus help to identify points of risk.

Summary

This chapter discussed common supply chain challenges in developing countries and their implications for packaging. Six categories of supply chain challenges were identified: infrastructure, organization and knowledge, logistics equipment, costs, climate and security.

The chapter highlighted that packaging needs to be better at protecting and communicating handling and stacking requirements as well as fragility concerns in a clear manner due to inadequate infrastructure for transportation and storage. Several challenges were discussed related to organization and knowledge of packaging and logistics, and their implications for packaging. The challenges include a lack of training programmes for logisticians and staff working with packaging, a low level of standardization in logistics, a low level of collaboration between companies, and the fact that supply chains have many stages. Many of these challenges affect the protective features of packaging, but they also affect handleability, stackability, labels with sufficient information and communication, and more. Logistics equipment is commonly inadequate which presents challenges to packaging protection, the facilitation of efficient handling, and more.

Next, the chapter discussed the need to have cost-efficient packaging solutions, which can result in the utilization of inexpensive and minimal packaging material, local sourcing of packaging material, and manual handling often being the preferable alternative. These effects put emphasis on controlling and enabling sufficient packaging protection and material quality, and that the packaging enables effective and efficient manual handling. Packaging is exposed to temperature and humidity challenges in many developing countries. To compensate for such challenges, packaging needs to have extra protection features. Finally, the chapter highlighted security in packed products related to theft and counterfeiting. Packaging can reduce both the risks of theft and counterfeiting by being hard to open, hard to identify orally, and by having traceability capabilities.

QUESTIONS

This section provides focused questions for each chapter. They will help you consolidate the knowledge you have gained from reading the book.

They will also useful for readers who would like to learn about packaging logistics but don't have time to read the book in its entirety.

They are available to download at **www.koganpage.com/packaginglogistics.**

Chapter 1 questions

DEFINING PACKAGING LOGISTICS

Describe the levels of a packaging system.

Describe the six basic functions of a packaging system.

Explain the difference between open and closed systems.

Describe the key characteristics of a systems approach.

Discuss at least two reasons why a systems approach is useful for packaging logistics.

Describe how supply chain integration is a key part of packaging logistics.

Discuss three examples of how packaging can support supply chain integration.

Define *transport packaging*, *industrial packaging*, *group packaging*, *shelf-ready packaging*, and *used packaging*.

Describe the meaning of trade-offs and interactions in packaging systems.

Chapter 2 questions

PACKAGING PERFORMANCE IN SUPPLY CHAINS

Explain how packaging can generate value.

Discuss and give examples of common trade-offs in a packaging system.

Explain why standardization can result in a lock in effect.

Describe what *standardization* of *packaging*, *transport*, *material handling*, and *ICT* refer to.

Give some examples of *packaging standardization* and *transport standardization.*

Explain four postponement and speculation strategies.

Discuss how packaging is affected by each strategy.

Chapter 3 questions

PACKAGING PERFORMANCE METHODOLOGY

Describe the four steps in the packaging performance methodology.

Give examples of product features and how they affect packaging.

Explain the process of mapping in terms of top-level mapping vs. detailed activity mapping.

Explain how top-level mapping and detailed activity mapping is related to zooming in and zooming out (see Chapter 1).

Discuss challenges in mapping the supply chain.

Explain how the packaging scorecard functions.

Explain how the interaction framework (in Figure 3.4) can be used.

Explain how packaging performance can be evaluated graphically.

Describe how to carry out the last step of the packaging performance methodology.

Explain organizational challenges to address in order to implement a new or modified packaging system.

Chapter 4 questions

MANAGING THE ENVIRONMENTAL IMPACT OF PACKAGING LOGISTICS

Explain the direct and indirect environmental impacts of packaging.

Explain how product characteristics and differences in supply chains affect the indirect environmental impacts of packaging.

Explain the role of packaging in environmentally responsible decision-making within *raw material acquisition, inbound logistics* and *marketing*.

Explain the trade-off between packaging material and product waste.

Describe an advantage of using less mixed materials.

Describe the environmental impact of three packaging features related to the *packaging material*.

Describe the environmental impact of three packaging features related to *logistics and transport efficiency*.

Describe the environmental impact of three packaging features related to the *product waste*.

Chapter 5 questions

MANAGING PACKAGING WASTE

Give examples of primary packaging waste, secondary packaging waste, and tertiary packaging waste.

Describe the characteristics and differences of household packaging waste and industrial packaging waste.

Explain where packaging waste for consumer goods occurs in a supply chain.

Describe two general differences for packaging waste in supply chains with a store and e-commerce supply chains with pick-up points.

Describe two general principles for managing packaging waste in a supply chain.

Discuss how to ensure that the storage capacity of packaging waste is sufficient in a production plant.

Chapter 6 questions

MANAGING ICT FEATURES OF PACKAGING

Define tracking and tracing.

Explain why packaging has a key role in track & trace.

Describe three organizational abilities that indicate high logistics service performance.

Give three examples of how packaging can affect logistics service performance.

Explain why barcodes are widely used.

Describe three decision criteria for choosing between barcodes and RFID technology, and describe also how each criterion influences the choice.

Explain why it may be difficult for an actor to get information from the whole supply chain.

Explain why information sharing of tracking data can be beneficial.

Explain four (out of eight) attributes to address in the design of a track & trace system for packaging.

Discuss three recommendations of how to implement track & trace capabilities through packaging in an inter-organizational group of companies.

Chapter 7 questions

REUSABLE PACKAGING SYSTEMS

Compare reusable to one-way packaging systems.

Describe the supply-chain characteristics that favour reusable packaging systems.

Explain why reusable packaging requires good management of reverse logistics.

Describe the main characteristics of the four basic types of reusable containers.

Choose and motivate a reusable container: It should facilitate efficient return transport, support manual handling in a warehouse, and contain a product with high hygienic requirements.

Explain the characteristics of a switch pool system.

Describe how a *transfer system*, a *depot system with booking*, and a *depot system with deposit* function in the control of reverse logistics.

Chapter 8 questions

INDUSTRIAL PACKAGING EVALUATION

Describe the main purposes of industrial packaging.

Discuss how to set system boundaries for evaluating costs and CO_2 emissions of industrial packaging.

Five factors should be included to calculate the impact of packaging on supply-chain costs. Explain the content of the *transport factor*, the *material handling factor*, and the *waste handling factor*. Describe the other two factors.

Explain how the volume efficiency of the packaging system affects the four CO_2 factors.

Give three examples of how an evaluation model for industrial packaging can support decision-making.

Explain how to compare the total supply-chain costs and CO_2 of one packaging system to another.

Discuss the proposed steps in the process of developing and implementing an evaluation model for industrial packaging.

Chapter 9 questions

DECISION SUPPORT FOR INDUSTRIAL PACKAGING

Explain why it is useful to understand the relative importance of different cost and CO_2 factors.

Discuss the impacts on total costs of using full pallets or less-than-full pallets.

Discuss how the material supply and production strategies can affect packaging efficiency.

Compare the pros and cons of using small or large packaging sizes.

Explain how filling up packages or pallets affect the different cost and CO_2 factors.

Give an example from the automotive industry of the need to include the context characteristics in calculating the impact of repacking on costs and CO_2.

Describe why scale effects should be analysed, and give an example from the automotive industry.

Chapter 10 questions

ORGANIZING PACKAGING DEVELOPMENT

Give one reason for designing and developing products and packaging simultaneously.

Describe four generic structures for organizing packaging development.

Explain the basic characteristics of a project-based and centralized packaging development organization.

Discuss how up-to-date packaging knowledge are affected by whether the packaging development organization is centralized or decentralized.

Explain the different approaches to capture conflicting packaging requirements in a function-based organization and in a project-based organization.

Discuss how packaging requirements from various stakeholders are included in a function-based organization and in a project-based organization.

Explain what it means that a function-based organization focuses on packaging and packaging development, while a project-based organization focuses on the packed product.

Give two reasons why concurrent product and packaging development can be beneficial.

Chapter 11 questions

PACKAGING INNOVATION CAPABILITY

Describe the difference between incremental and radical innovation.

Explain *incremental*, *modular*, *architectural*, and *radical* packaging concept innovation.

Describe the seven stages in the process of generating packaging concepts.

Describe two areas of idea generation and discuss their impact on for new packaging concepts.

Explain two prerequisites in the concept development stage to create high-quality packaging concepts.

Explain the feasibility test in the process of generating packaging concepts.

Discuss how to assess the quality of packaging innovation planning.

The fourth process of the packaging innovation capability is packaging development. Discuss how to assess the quality of this process.

Give an example of a high-quality process indicator for the process 'transfer new packaging designs to manufacturing and distribution'.

Which are the other processes that determine the quality of the packaging innovation capability.

Chapter 12 questions

MANAGING PACKAGING LOGISTICS IN E-COMMERCE CHANNELS

Explain how the last mile changes in e-commerce with home delivery compared to sales in a store, and how this affects packaging.

Describe a common practice for packaging solutions in e-commerce.

Describe three alternative ways of organizing pick-up points.

Describe three effects on packaging when they are supplied to a pick-up point.

Describe potential effects of e-commerce on retail, and their implications for the packaging system.

Explain how e-commerce can affect the packaging features apportionment and stackability.

Give three other examples (than apportionment and stackability) of packaging features that are should be addressed differently in e-commerce than in conventional trade.

Chapter 13 questions

MANAGING PACKAGING LOGISTICS IN DEVELOPING COUNTRIES

Discuss the number of supply chain stages and the characteristics of material handling in developing countries. Discuss the implications of this for packaging.

Describe six categories of supply chain challenges in developing countries.

Discuss how inadequate infrastructure affects packaging.

Explain the problem with poor building quality and its impact on packaging.

Give two examples of challenges related to 'organization and knowledge' in developing countries, and how these challenges affect packaging.

Explain why a low level of logistics standardization is inefficient.

Give two examples of challenges related to costs in developing countries, and how these challenges affect packaging.

REFERENCES

Ballou, RH (2004) *Business Logistics/Supply Chain Management: Planning, organizing, and controlling the supply chain*, 5th edn, Prentice Hall, Upper Saddle River, NJ

Bechini, A, Cimino, MGCA, Marcelloni, F and Tomasi, A (2008) Patterns and technologies for enabling supply chain traceability through collaborative e-business, *Information and Software Technology*, 50(4), pp 342–59

Bond, M, Meacham, T, Bhunnoo, R and Benton, TG (2013) Food waste within global food systems: A Global Food Security report (www.foodsecurity.ac.uk)

Bowersox, DJ, Closs, DJ and Cooper, MB (2002) *Supply Chain Logistics Management*, First International edn, McGraw-Hill/Irwin, New York

Caraballo, V and Westergren, B (2012) Redefinition of pallet configurations, Master's thesis, Faculty of Engineering, Lund University, Lund, Sweden

Carlsson, S, Pålsson, H, Wallström, H and Johansson, M (2015) A model for packaging systems evaluation from a sustainability perspective in the automotive industry, Final report of a project within Sustainable Production Technology (VINNOVA), Stockholm, Sweden

Chiesa, V, Coughlan, P and Voss, CA (1994) Development of a technical innovation audit, *Journal of Product Innovation Management*, 13(2), pp 105–36

Council of Supply Chain Management Professionals (CSCMP, 2018) Definition of Logistics Management, https://cscmp.org/imis0/CSCMP/Educate/SCM_Definitions_and_Glossary_of_Terms/CS CMP/Educate/SCM_Definitions_and_Glossary_of_Terms.aspx (accessed 7 April 2018)

Davenport, TH and Short, JE (1990) The new industrial engineering: Information technology and business process redesign, *Sloan Management Review*, 31(4), pp 1–31

Doherty, S and Hoyle, S (2009) Supply chain decarbonization: The role of logistics and transport in reducing supply chain carbon emissions, World Economic Forum, Geneva

du Preez, ND, Louw, L and Essmann, H (2009) An innovation process model for improving innovation capability, *Journal of High Technology Management Research*, pp 1–24

ECR Europe (2007) *Shelf Ready Packaging: Addressing the challenge: A comprehensive guide for a collaborative approach*, http://www.ecr.digital/wp_contents/uploads/2016/09/ECR-Bericht_Shelf_Ready_Packaging.pdf (accessed 31 January 2018), pp 1–51

European Agency for Safety and Health at Work (2018), Dangerous substances in packaging, link (accessed 11 January 2018), https://osha.europa.eu/en/themes/dangerous-substances/clp-classification-labelling-and-packaging-of-substances-and-mixtures

Eurostat (2017) *Packaging Waste Statistics*, http://ec.europa.eu/eurostat/statistics-explained/index.php?title=Packaging_waste_statistics&oldid=243064 (accessed 11 April 2018)

Hanssen, OJ (1998) Environmental impacts of product systems in a life cycle perspective: A survey of five product types based on life cycle assessments studies, *Journal of Cleaner Production*, 6(3/4), pp 299–311

Hellström, D and Saghir, M (2007) Packaging and logistics interactions in retail supply chains, *Packaging Technology and Science*, 20, pp 197–216

Henderson, RM and Clark, KB (1990) Architectural innovation: The reconfiguration of existing product technologies and the failure of established firms, *Administrative Science Quarterly*, 35(1), pp 9–30

IEA (2017) Transport sector indicators, https://www.iea.org/Day_2_Indicators_training_Exercice1_ANSWERS.xls (accessed 21 August 2017)

INCPEN (2009) Table for one: The energy cost to feed one person, http://www.incpen.org/docs/TableForOne.pdf (accessed 11 January 2018)

ITF (2010) Reducing transport greenhouse gas emissions: Trends and data 2010, International Transport Forum, OECD report

Johansson, O and Pålsson, H (2009) The impact of Auto-ID on logistics performance: A benchmarking survey of Swedish manufacturing industries, *Benchmarking: An International Journal*, 16(4), pp 504–22

Klevås, J (2005) Organization of packaging resources at a product-developing company, *International Journal of Physical Distribution & Logistics Management*, 35(2), pp 116–131

Klevås, J, Johnsson, M and Jönson, G (2005) A packaging redesign project at IKEA, Department of Design Sciences, Lund University, pp 1–10

Kovács, G and Spens, K (2009) Identifying challenges in humanitarian logistics, *International Journal of Physical Distribution & Logistics Management*, 39(6), pp 506–28

Kroon, L and Vrijens, G (1995) Returnable containers: An example of reverse logistics, *International Journal of Physical Distribution & Logistics Management*, 25(2), pp 56–68

Livingstone, S and Sparks, L (1994) The new German packaging laws: Effects on firms exporting to Germany, *International Journal of Physical Distribution & Logistics Management*, 24(7), pp 15–25

Lockamy III, A (1995) A conceptual framework for assessing strategic packaging decisions, *The International Journal of Logistics Management*, 6(1), pp 51–60

Lützebauer, M (1993) Mehrwegsysteme für Transportverpackungen: Marktreport, 2nd edn, Deutscher Fachverlag GmbH, Frankfurt am Main, ISBN: 3871504505

Mele, C, Pels, J and Polese, F (2010) A brief review of systems theories and their managerial applications, *Service Science*, 2(1–2), pp 126–35

Min, H, Ko, HJ, Lim, YK, Park, JW and Cho, YK (2014) Challenges and opportunities for logistics standardisation in Asia-Pacific countries: A descriptive case study, *International Journal of Logistics Systems and Management*, 17(3), pp 357–80

Molina-Besch, K and Pålsson, H (2013) Packaging for eco-efficient supply chains: Exploring factors in the packaging development process, in *Nofoma Proceddings*, pp 1–16

Molina-Besch, K and Pålsson, H (2016) A supply chain perspective on green packaging development: Theory versus practice, *Packaging and Technology and Science*, 29, pp 45–63

O'Leary-Kelly, SW and Flores, BE (2002) The integration of manufacturing and marketing/sales decisions: Impact on organizational performance, *Journal of Operations Management*, 20(3), pp 221–40

Olsmats, C and Dominic, C (2003) Packaging scorecard: A packaging performance evaluation method, *Packaging Technology and Science*, 16, pp 9–14

Pagh, JD and Cooper, MC (1998) Supply chain postponement and speculation strategies: How to choose the right strategy, *Journal of Business Logistics*, 19(2), pp 13–33

Paine, F (1981) *Fundamentals of Packaging*, Brookside Press, Leicester, UK

Parfitt, J, Barthel, M and Macnaughton, S (2010) Food waste within food supply chains: Quantification and potential for change to 2050, *Philosophical Transactions of the Royal Society B*, 365, pp 3065–081

Pålsson, H (2007) Participant observation in logistics research: Experiences from an RFID implementation study, *International Journal of Physical Distribution & Logistics Management*, 37(2), pp 148–163

Pålsson, H and Johansson, O (2009) Supply chain integration obtained through uniquely labelled goods: A survey of Swedish manufacturing industries, *International Journal of Physical Distribution & Logistics Management*, 39(1), pp 28–46

Pålsson, H, Finnsgård, C and Wänström, C (2013) Selection of packaging systems in supply chains from a sustainability perspective: The case of Volvo, *Packaging Technology and Science*, 26(5), pp 289–310

Pålsson, H, Pettersson, F and Winslott Hiselius, L (2017) Energy consumption in e-commerce versus conventional trade channels: Insights into packaging, the last mile, unsold products and product returns, *Journal of Cleaner Production*

Pålsson, H, Wallström, H and Johansson, O (2014) Returnable vs. one-way packaging: Variables affecting supply chain cost and CO_2 emissions, Proceedings of EUROMA, Palermo, Italy

Robertson, GL (1990) Good and bad packaging: Who decides, *International Journal of Physical Distribution and Logistics Management*, pp 37–40

Saruchera, F (2017) Rail freight transportation concerns of developing economies: A Namibian perspective, *Journal of Transport and Supply Chain Management,* 11, pp 1–9

Siikavirta, H, Punakivi, M, Ka, M and Linnanen, L (2003) Effects of e-commerce on greenhouse gas emissions: A case study of grocery home delivery, *Journal of Industrial Ecology*, 6(2), pp 83–97

Sivaraman, D, Pacca, S, Mueller, K and Lin, J (2007) Comparative energy, environmental, and economic analysis of traditional and e-commerce DVD rental networks, *Journal of Industrial Ecology*, 11(3), pp 77–91

Souhrabpour, V, Hellström, D and Jahre, M (2012) Packaging in developing countries: Identifying supply chain needs, *Journal of Humanitarian Logistics and Supply Chain Management*, 2(2), pp 183–205

Stank, T, Keller, SB and Daugherty, PJ (2001) Supply chain collaboration and logistical service performance, *Journal of Business Logistics*, 22(1), pp 29–48

Stefansson, G and Tilanus, B (2001) Tracking and tracing: Principles and practice, *International Journal of Services Technology and Management,* 2(3), pp 187–206

Stenmarck, Å, Jensen, C, Quested, T and Moates, G (contributing partners: Buksti, M, Cseh, B, Juul, S, Parry, A, Politano, A, Redlingshofer, B, Scherhaufer, S, Silvennoinen, K, Soethoudt, H, Zübert, C, Östergren, K) (2016) Estimates of European food waste levels, FUSIONS: Reducing food waste through social innovation, ISBN 978-91-88319-01-2, (http://www.eu-fusions.org/phocadownload/Publications/Estimates%20of%20European%20food%20waste%20levels.pdf), pp 1–80

Stevens, GC (1989) Integrating the supply chain, *International Journal of Physical Distribution & Logistics Management*, 19(8), pp 3–8

Stock, JR (1998) *Development and Implementation of Reverse Logistics Programs*, Council of Logistics Management, Oakbrook, IL

Suhaimi, MS (2017) Evaluation of the corrugated box strength performance in supply chains: A case study of Duni AB, Master's thesis, Faculty of Engineering, Lund University, Lund, Sweden

Swedish Regulation 2014:1073 (2014), Förordning om producentansvar för förpackningar, https://www.riksdagen.se/sv/dokument-lagar/dokument/svensk-forfattningssamling/forordning-20141073-om-producentansvar-for_sfs-2014-1073 (accessed 11 April 2018)

Twede, D and Clarke, RH (2004) Supply chain issues in reusable packaging, *Journal of Marketing Channels*, 12(1), pp 7–26

Twede, D, Clarke, RH and Tait, J (2000) Packaging postponement: A global packaging strategy, *Packaging Technology and Science*, 13(3), pp 105–115

van der Mescht, J (2006) Revisiting the road versus rail debate, *25th Southern African Transport Conference*, 1(July), pp 485–94

Williams, H and Wikström, F (2010) Environmental impact of packaging and food losses in a life cycle perspective: A comparative analysis of five food items, *Journal of Cleaner Production*, 19, pp 43–48

Williams, H, Wikström, F, Otterbring, T, Löfgren, M and Gustafsson, A (2012) Reasons for household food waste with special attention to packaging, *Journal of Cleaner Production*, 24(1), pp 141–48

Wu, H-J and Dunn, SC (1995) Environmentally responsible logistics systems, *International Journal of Physical Distribution & Logistics Management*, 25(2), pp 20–38

Zhang, Q, Segerstedt, A, Tsao, Y-C and Liu, B (2015) Returnable packaging management in automotive parts logistics: Dedicated mode and shared mode, *International Journal of Production Economics*, 168, pp 234–44

FURTHER READING

Accorsi, R, Cascini, A, Cholette, S, Manzini, R and Mora, C (2014) Economic and environmental assessment of reusable plastic containers: A food catering supply chain case study, *International Journal of Production Economics* Elsevier, 152, pp 88–101

Early, C, Kidman, T, Menvielle, M, Geyer, R and McMullan, R (2009) Informing packaging design decisions at Toyota Motor sales using life cycle assessment and costing, *Journal of Industrial Ecology*, 13(4), pp 592–606

García-Arca, J Prado-Prado, J C and Gonzalez-Portela Garrido, A T (2014) Packaging logistics: promoting sustainable efficiency in supply chains, *International Journal of Physical Distribution & Logistics Management*, 44(4), pp 325–346

Göransson, M, Jevinger, & Nilsson, J (2018) Shelf-life variations in pallet unit loads during perishable food supply chain distribution *Food Control*, 84, 552–560

Grönman, K, Soukka, R, Järvi-Kääriäinen, T, Katajajuuri, J-M, Kuisma, M, Koivupuro, H-K, Ollila, M, Pitkänen, M, Miettinen, O, Silvenius, F, Thun, R, Wessman, H and Linnanen, L (2013) Framework for Sustainable Food Packaging Design, *Packaging and Technology and Science*, 26, pp 187–200

Hellström, D and Pålsson, H (2011) 16 RFID-Enabled Visibility in a Dairy Distribution Network, I M Bourlakis, I Vlachos, & V Zeimpekis (Eds), *Intelligent Agrifood Chains and Networks*, pp 267–280 Wiley-Blackwell

Hellström, D and Nilsson, F (2011), Logistics-driven packaging innovation: a case study at IKEA, International Journal of Retail & Distribution Management, Vol 39 Issue 9, pp 638–657

Jahre, M and Hatteland, C J (2004) Packages and physical distribution: Implications for integration and standardisation, *International Journal of Physical Distribution & Logistics Management*, 34(2), pp 123–139

Kye, D, Lee, J and Lee, K (2013) The perceived impact of packaging logistics on the efficiency of freight transportation (EOT), *International Journal of Physical Distribution & Logistics Management*, 43(8), pp 707–720

Lai, J, A Harjati, L McGinnis, C Zhou, and T Guldberg 2008 An economic and environmental framework for analyzing globally sourced auto parts packaging system journal of cleaner production 16 (15): 1632–1646

Molina-Besch, K, and H Pålsson 2014a Packaging for eco-efficient supply chains: why logistics should get involved in the packaging development process in *Sustainable Logistics* edited by C Macharis, S Melo, J Woxenius, and T van Lier, pp 137–163, Emerald Group Publishing Limited (Transport and Sustainability)

Mollenkopf, D, Closs, D, Twede, D, Lee, S and Burgess, G (2005) Assessing the viability of reusable packaging: a relative cost approach, *Journal of Business Logistics*, 26(1), pp 169–197

Pålsson, H and Hellström, D (2016) Packaging logistics in supply chain practice: Current state, trade-offs and improvement potential, *International Journal of Logistics Research and Applications*, 19(5), pp 351–368

Sparks, L, Gustafsson, K, Jönson, G & Smith, D (2006) *Retailing Logistics and Fresh Food Packaging: Managing Change in the Supply Chain* Kogan Page

Svanes, E, M Vold, H Möller, M K Pettersen, H Larsen, and O-J Hanssen 2010 Sustainable Packaging Design: A Holistic Methodology for Packaging Design Packaging Technology and Science 23 (3): 161–175

Vakulenko, Y, Hellström, D & Hjort, K (2017) What's in the parcel locker? Exploring customer value in e-commerce last mile delivery *Journal of Business Research*

Vergese, K, Lewis, H, Lockrey, S and Williams, H (2015) Packagings role in minimizing food loss and waste across the supply chain, *Packaging and Technology and Science*, 23(May), pp 253–266

GLOSSARY

apportionment (packaging function) reducing large-scale and high-volume production to manageable sizes

backhaul management managing the return transport of a truck back to its origin with the purpose of hauling cargo back on this trip as well

backward integration a form of vertical supply chain integration in the direction away from the end customer, eg when a manufacturer integrates with a supplier

barcodes codes designed to be optically scanned. A code represents an object's data. It consists of a group of bars and spaces, and sometimes digits

blister packaging pre-formed plastic packaging with a pocket or a cavity made from a formable web. Used for pharmaceuticals, food, electronics, toys and other consumer goods

bulk packaging (same as distribution, industrial and transport packaging) packaging that facilitates handling, transport and storage of a number of primary packages in order to provide efficient production and distribution, as well as to prevent physical handling and damage during transportation

centralized organization an organization where the decision-making authority is consolidated to the top management in the hierarchical structure. Typically, a few senior managers in the head office have the major responsibilities and power

closed system a system isolated from the external environment without any interactions outside the system boundaries

communication (packaging function) packaging identification in the supply chain and the provision of product information

consumer packaging (same as primary packaging) Packaging which is in contact with the product, and the one that consumers usually bring home

containment (packaging function) holding and maintaining the content

convenience (packaging function) simplifying the use of products

cumulative energy demand (CED) the total primary energy consumed when producing a product or a package

decentralized organization an organization where the decision-making authority is spread to lower hierarchical levels in the organization. Typically, the responsibilities and power are spread to lower level managers in sub-units, which can be located away from the head office

display packaging (same as group packaging with an emphasis on display features) packaging that is created to facilitate protection, display, handling, and/or transportation of a number of primary packages

distribution packaging (same as bulk, industrial and transport packaging) packaging that facilitates handling, transport and storage of a number of primary packages in order to provide efficient production and distribution, as well as to prevent physical handling and damage during transportation

economic order quantity (EOQ) the number of units that a company should order to minimize the total costs of inventory associated with the purchase, delivery and storage costs of a unit

economies of scale the scale of operations leads to a reduced cost *per unit*, because of operational efficiencies and synergies that divide fixed costs and overhead costs to more units

external integration integration between companies, both in the direction toward and away from the end customer

forward integration a form of vertical supply chain integration towards the end customer, eg when a manufacturer integrates with a customer

group packaging packaging that is created to facilitate protection, display, handling, and/or transportation of a number of primary packages

horizontal integration integration between competitors

inbound logistics the processes related to incoming goods, such as transportation, receiving, storing, and distributing raw material for use in production

industrial packaging (same as bulk, distribution and transport packaging) packaging that facilitates handling, transport and storage of a number of primary packages in order to provide efficient production and distribution, as well as to prevent physical handling and damage during transportation

internal integration integration between functions within a company

life cycle assessment (LCA) an analytical approach that aims to capture the overall environmental impacts associated with all the stages of an entity's life (usually a product's) from raw material acquisition, through production and use to waste management

material handling movement and handling of goods or materials within and between storage areas, within a manufacturing facility, within a warehouse etc. It includes forklift operations, loading and unloading, palletizing and de-palletizing, packing and repacking etc. The operations can be manual, semi-automated and fully automated

modularization of packaging systems packaging systems are seen as modular when they can be decomposed into a number of packaging units (eg primary and secondary packages) that may be mixed and matched with other modularized packages in a variety of configurations to form other packaging systems

near-field communication (NFC) communication protocols that enable two electronic devices in close proximity to communicate

normalization adjusting values on different scales to a common scale

one-way packaging designed to be used once, eg from the point of filling to the point of sale or the point of consumption. After the end of use, one-way packaging becomes waste

open system an open system has external interactions outside the system boundaries. Exchanges of information and material with the surrounding environment needs to be considered

outbound logistics the processes related to outgoing goods, such as storage, transportation, and distribution of goods to customers

packaging concept a platform that determines certain packaging attributes, such as the graphical design, material or material performance, shape and labelling style, but do not have to include all of these

packaging system a packaging system consists of primary, secondary and tertiary packaging, and their interrelations

postponement to make the earlier stages of a supply chain more efficient, activities can be postponed until more information is available or when diversity is needed

primary packaging packaging which is in contact with the product, and the one that consumers usually bring home

process mapping mapping and describing the flow of work, goods, information etc in a series of events from point of origin to the end result. A clear and detailed process map provides an understanding of how a process and its parallel processes work. It highlights the characteristics of the current process and its improvements potential

protection (packaging function) safeguard the content

quick response (QR) codes two-dimensional codes designed to be optically scanned. A code represents data of an object. QR codes can be read by an imaging device such as a mobile phone camera

radio-frequency identification (RFID) technology a wireless communication technology that uses electromagnetic fields to uniquely identify and track tags attached to objects without a requirement on line-of-sight from the reader

replenishment the operations that makes the stock full again in order to avoid stock-out. Can refer to replenishing inventory in warehouses, in retail stores, on shelves etc

reusable packaging designed to be reused without weakening its protective function. Reusable packaging requires reverse logistics and the management of this process

reverse logistics the role of logistics in product returns, source reduction, recycling, materials substitution, reuse of materials, waste disposal and refurbishing, repair and remanufacturing (Stock, 1998, p. 2)

sales packaging (same as primary packaging) packaging which is in contact with the product, and the one that consumers usually bring home

secondary packaging secondary packaging contains a certain number of primary packages

shelf-ready packaging (SRP) packaging created to facilitate protection, display, handling, and/or transportation of a number of primary packages with a special emphasis on the design fitting in retail stores

standardization aims to reduce complexity, target selected activities, protect the interest of majorities, persist over time, but be subject to review and change when needed

sub-optimization optimization of a part of a system or a part of an organization rather than of the system or organization as a whole

supply chain integration (SCI) coordinating and integrating separate parts of a supply chain into a cohesive whole. The underlying logic is to synchronize and integrate processes and activities with a focus on the end customer offer

system a group of individual parts that work together to form a unified whole. It is a way to examine a phenomenon or linked parts as one unit, instead of as individual parts. A system is characterized by the whole system not having to be equal to the sum of its parts, as these parts are connected to each other to form a whole. An essential characteristic is the focus on the interactions between the parts. A system contains a clear system boundary

tertiary packaging tertiary packaging contains a certain number of secondary packages

total cost analysis analysing how a logistics decision affects the total costs of logistics, which consist of a range of costs associated with logistics, such as material handling, transport and warehousing costs, as well as inventory carrying, administration and order processing costs. A logistics decision can reduce or increase many of these costs

tracing the capability to *find* a packed product between the point of origin and the point of consumption

tracking the capability to *follow* a packed product on its way from point of origin to point of consumption

trade-off balancing several factors, which are not attainable at the same time. With trade-offs some desirable outcomes are reduced or ignored in exchange for other desirable outcomes with the aim of maximizing the total effectiveness and efficiency under given circumstances

transport packaging (same as bulk, distribution and industrial packaging) packaging that facilitates handling, transport and storage of a number of primary packages in order to provide efficient production and distribution, as well as to prevent physical handling and damage during transportation

unitization (packaging function) unitize a number of primary packages to fit in a secondary package, and unitize a number of secondary packages to fit in a tertiary package

used packaging packaging or packaging material remaining after the removal of the product it contained

vertical integration integration with suppliers and customers

waste fraction when waste is grouped according to its properties, each group is called a waste fraction

INDEX

KOGAN PAGE
LOGISTICS

Cutting-edge insights from leading thinkers in academia and industry

Decarbonizing Logistics outlines the nature and extent of the challenge we face in trying to achieve deep reductions in greenhouse gas emissions from logistical activities. It makes a detailed assessment of the available options, including restructuring supply chains, shifting freight to lower-carbon transport modes and transforming energy use in the logistics sector. The options are examined from technological and managerial standpoints for all the main freight transport modes.

Based on an up-to-date review of almost 600 publications and containing new analytical frameworks and research results, *Decarbonizing Logistics* is the first book to provide a global, multidisciplinary perspective on the subject. It is written by one of the foremost specialists in the field.

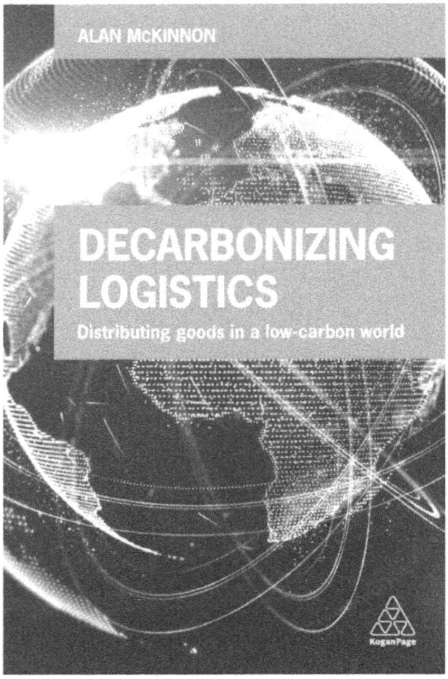

Alan McKinnon is Professor of Logistics at Kühne Logistics University, Hamburg. He has been researching and teaching freight transport and logistics for almost forty years and has published extensively in journals and books. He was a member of the European Commission's High Level Group on Logistics, Chairman of the World Economic Forum's Logistics and Supply Chain Industry Council and a lead author of the transport chapter of the Intergovernmental Panel on Climate Change's Fifth Assessment Report. He has spent many years researching the links between logistics and climate change and been an adviser to governments, international organizations and companies.

DECARBONIZING LOGISTICS

www.koganpage.com/DL

CPSIA information can be obtained
at www.ICGtesting.com
Printed in the USA
LVHW01s1801031018
592283LV00013B/188/P

9 780749 481704